# Medical Genetics for the MRCOG and Beyond

*Second edition*

# Medical Genetics for the MRCOG and Beyond

## Second edition

Edward S. Tobias and J. Michael Connor

**CAMBRIDGE**
**UNIVERSITY PRESS**

University Printing House, Cambridge CB2 8BS, United Kingdom

Published in the United States of America by Cambridge University Press, New York

Cambridge University Press is part of the University of Cambridge.

It furthers the University's mission by disseminating knowledge in the pursuit of education, learning and research at the highest international levels of excellence.

www.cambridge.org

Information on this title: www.cambridge.org/9781107661301



First published 2014

Printed in Spain by Grafos SA, Arte sobre papel

*A catalog record for this publication is available from the British Library*

*Library of Congress Cataloging-in-Publication Data*
Tobias, Edward S., author.
Medical genetics for the MRCOG and beyond / Edward Tobias and J. Michael Connor. – Second edition.
p. ; cm.
Preceded by: Medical genetics for the MRCOG and beyond / J. Michael Connor. 2005.
Includes bibliographical references and index.
ISBN 978-1-107-66130-1 (Paperback)
I. Connor, J. M. (James Michael), 1951- author. II. Royal College of Obstetricians and Gynaecologists (Great Britain), publisher. III. Title.
[DNLM: 1. Genetics, Medical. 2. Pregnancy Complications. WQ 240]
RB155.6
616′.042–dc23 2013045691

ISBN 978-1-107-66130-1 Paperback



---

# Contents

# Abbreviations

| | |
|---|---|
| **A** | adenine |
| **aCGH** | array comparative genomic hybridisation |
| **α-FP** | alphafetoprotein |
| **ARMS** | amplification refractory mutation system |
| **bp** | base pair |
| ***BRCA1*** | breast cancer type 1 gene |
| **C** | cytosine (or consultand in a pedigree diagram) |
| **cffDNA** | cell-free fetal DNA |
| **CFTR** | cystic fibrosis transmembrane conductance regulator gene |
| **CPK** | creatine phosphokinase |
| **CVS** | chorionic villus sampling |
| **DMD** | Duchenne muscular dystrophy |
| ***DMPK*** | dystrophia myotonica-protein kinase gene |
| **DNA** | deoxyribonucleic acid |
| **EDTA** | ethylenediamine tetra-acetic acid |
| **FβhCG** | free beta human chorionic gonadotrophin |
| **FISH** | fluorescence *in situ* hybridisation |
| ***FMR1*** | fragile site mental retardation 1 gene |
| **FXTAS** | fragile X tremor/ataxia syndrome |
| **G** | guanine |
| **Gy** | gray |
| **hCG** | human chorionic gonadotrophin |
| **HNPCC** | hereditary nonpolyposis colorectal cancer |
| ***HPRT*** | hypoxanthine phosphoribosyltransferase gene |
| ***IDUA*** | alpha-L-iduronidase gene |
| **IRT** | immunoreactive trypsinogen |
| **kb** | kilobase |

| | |
|---|---|
| ***L1CAM*** | L1 cell adhesion molecule gene |
| **Mb** | megabase |
| **MLPA** | multiplex ligation-dependent probe amplification |
| **M/M** | mutant/mutant |
| **MOM** | multiples of the median |
| **MSAFP** | maternal serum alphafetoprotein |
| **N/M** | normal/mutant |
| **NGS** | next-generation sequencing |
| **NIPD** | noninvasive prenatal diagnosis |
| **PAPP-A** | pregnancy-associated plasma protein A |
| **PCR** | polymerase chain reaction |
| **PGD** | preimplantation genetic diagnosis |
| **QF-PCR** | quantitative fluorescent polymerase chain reaction |
| **rads** | radiation absorbed doses |
| **T** | thymine |
| **TP63** | tumour protein p63 |

# Glossary

| | |
|---|---|
| allele | alternative forms of a gene at the same locus |
| array comparative genomic hybridisation (aCGH) | detection method for DNA duplications or deletions by competitive hybridisation, using a microarray of known mapped sequences and fluorescently labelled control and test DNA |
| autosomal dominant inheritance | mutation in one member of an autosomal gene pair results in disease |
| autosomal recessive inheritance | mutation in both members of an autosomal gene pair is necessary for disease to occur |
| autosome | chromosomes numbers 1 to 22 inclusive |
| balanced translocation | transfer of chromosomal material between chromosomes with no overall gain or loss and hence no clinical effect |
| base pair | unit of length of DNA of one set of paired bases (AT or GC) |
| carrier | a person with one mutation in an autosomal or X chromosomal gene pair who shows recessive inheritance (i.e. no clinical effect unless both members of the gene pair are mutated) |
| centromere | a constricted area of the chromosome that divides it into short and long arms |
| chromosome disorder | any abnormality of chromosome number or structure visible under the light microscope |
| codon | three consecutive bases in DNA (or RNA) that specify an amino acid |

| | |
|---|---|
| concordance | likelihood that both (e.g. twins) will be affected or unaffected |
| congenital | present at birth |
| consanguineous | mating between individuals who share at least one common ancestor |
| consultand | a person requesting genetic counselling |
| deletion | loss of chromosomal material |
| diagnostic test | a test that confirms or refutes a diagnosis |
| dizygotic twins | twins which arise from the fertilisation of two separate eggs |
| dominant | a trait expressed in the heterozygote |
| empiric risk | recurrence risk based on experience rather than calculation |
| false negative rate | proportion of affected cases missed by a screening test |
| first-degree relatives | immediate relatives who have one half of their genes in common (e.g. parent and child or brother and sister) |
| fluorescence *in situ* hybridisation | the use of a fluorescently labelled DNA probe to bind to specific chromosomal region of interest |
| gene | a segment of DNA that codes for a functional product (e.g. a protein) |
| gene probe | a labelled segment of DNA that can be used to find its matching segment among a mixture of DNA fragments |
| genetic counselling | communication of information and advice about inherited disorders |
| genetic heterogeneity | genetic mimicry where mutations in different genes can produce a similar clinical picture |
| genomic imprinting | parent-specific expression or repression of genes in offspring |
| genotype | the genetic make-up of an individual |
| gonadal mosaic | a person with a mixture of cells in their gonad, some with a mutation and some without |

| | |
|---|---|
| heterozygous | a person with a gene pair who has one mutant and one normal gene |
| homozygous | a person with a gene pair in which both copies of the gene are mutant or normal |
| independent risks | risks where the outcome of one event has no influence on the outcome of the other (e.g. if two coins are tossed heads or tails may occur for either and the result for one does not influence the other) |
| karyotype | the chromosomal make-up of an individual |
| kilobase | a unit of length of DNA of 1000 base pairs |
| length mutation | a type of DNA change where the DNA sequence is increased or decreased in size |
| locus | the location of a gene on a chromosome |
| megabase | a unit of length of DNA of 1 000 000 base pairs |
| meiosis | reduction cell division that occurs in the gonads in the production of eggs and sperm |
| microdeletion | a chromosomal deletion that is at or below the limit of resolution using a light microscope |
| mitosis | normal cell division that results in daughter cells with an identical genetic complement |
| monozygotic twins | twins which result from the early division of a single fertilised egg into two embryos |
| mosaic | an individual with cells with two or more genetic constitutions |
| multifactorial inheritance | conditions arising from the interaction of multiple genes and environmental factors |
| mutation | alteration of genetic material |

| | |
|---|---|
| mutational heterogeneity | different mutations in a particular gene may cause the same disease |
| mutually exclusive risks | risks where one outcome of an event precludes another outcome (e.g. a single tossed coin can result in heads or tails but not both) |
| nonpenetrance | no signs or symptoms in an individual who has inherited an autosomal dominant trait |
| phenotype | the clinical features of an individual |
| point mutation | a type of DNA change where a single base is replaced with another base |
| polymerase chain reaction | a technique for amplification of a target segment of DNA |
| polymorphism | a common DNA or chromosomal variant (present in at least 1% of the population) |
| proband | the individual who draws medical attention to the family |
| recessive | a trait that is expressed only in homozygotes |
| satellite stalks | the short arms of chromosomes 13, 14, 15, 21 and 22 |
| screening test | a test that divides a population according to risk for a condition; those at high risk are then offered a diagnostic test |
| second-degree relatives | close relatives with one-quarter of their genes in common (e.g. grandparent and grandchild or nephew/niece and aunt/uncle) |
| sensitivity | the proportion of cases detected by a screening test |
| sibship | a family group of brothers and/or sisters |
| somatic cell disorders | genetic conditions that arise after conception from accumulation of genetic mutations in a cell or group of cells |
| somatic mosaic | a person with a mixture of cells, some with a mutation and some without |

| | |
|---|---|
| specificity | the proportion of the unaffected population included by a screening test in the high-risk group (also called the false positive rate) |
| syndrome | a nonrandom combination of clinical features |
| telomere | the ends of the short and long arms of the chromosomes |
| third-degree relatives | more distant relatives who share one-eighth of their genes (e.g. first cousins) |
| trait | any gene-determined characteristic |
| translocation | the transfer of chromosomal material between chromosomes |
| triploidy | an extra half set of chromosomes resulting in 69 in total |
| trisomy | an extra copy of a chromosome resulting in 47 in total |
| variable expressivity or expression | variation in clinical effects of an autosomal dominant trait |
| X-linked recessive inheritance | disease due to mutations in genes on the X-chromosome; males with only one X are affected if that copy is mutant whereas females with two X chromosomes are usually unaffected if only one copy is mutant |

# Preface

There is a long history of successful interaction between obstetrics and gynaecology and medical genetics. Initially, most applications related to obstetrics, especially with the use of prenatal diagnosis and screening but, more recently, the growth has been in applications related to gynaecology, especially in relation to gynaecological malignancies.

However, despite this long history, there is a widespread misconception that genetics is a difficult subject to understand. This book thus aims to dispel this misconception as well as providing a revision aid for the MRCOG candidate. The first section covers basic principles. The second section outlines the more common situations where obstetrics and gynaecology and medical genetics interact and the third section contains real-life clinical case scenarios. These scenarios have been selected to represent typical problems and to highlight areas that, if mismanaged, could (and did, in many of these cases) lead to medico-legal consequences.

The book discusses the uses of the latest techniques, such as 'next-generation sequencing', quantitative fluorescent polymerase chain reaction (QF-PCR), array comparative genomic hybridisation (aCGH), preimplantation genetic diagnosis (PGD) and the recently introduced analysis of free fetal DNA in the maternal circulation for noninvasive prenatal diagnosis (NIPD). In addition, the increasing importance of online databases is reflected in the greatly expanded section (Appendix 1) that outlines the online medical genetic resources, which are most useful and appropriate for different purposes and provides their web addresses. An accompanying online guide (www.essentialmedgen.com) provides the reader with links to these databases from a single website in addition to news of the latest developments in the field.

Edward S. Tobias
J. Michael Connor

# Acknowledgements

We wish to thank all of the people who have contributed to, or influenced, the production of this book. These include all of our colleagues in Medical Genetics in Glasgow. We also wish to thank, in particular, Jim Colgan, Louise Brown, Alexander Cooke, Stuart Imrie, Gordon Lowther and Nicola Williams.

We are most obliged to Alexander Fletcher for his help in regularly checking all the live links on the accompanying website at www.essentialmedgen.com.

The authors are indebted to the editorial and production teams involved, including Jane Moody, Claire Dunn, Nicholas Dunton, Joanna Chamberlin and Rachel Cox.

EST would like to express his enormous gratitude to his wife, his family and his friends for their continuous support and understanding while he worked on the manuscript.

We are very grateful to the patients and their families and also to the following for their permission to reproduce these figures:

Figs. 1.28 and 1.29; front cover image: Norma Morrison

Figs. 1.1-1.8, 1.15, 1.32, 2.1, 2.14, 2.15 and 3.7: Jim Colgan

Fig. 1.23: Alexander Cooke

Figs. 1.24 and 1.25: Jim Kelly and Alexander Cooke

Fig. 1.30: Catherine McConnell

Fig. 1.31: Nicola Williams

Fig. 3.20: Margo Whiteford

We would also like to thank the curators of the National Centre for Biotechnology Information (NCBI) and Online Mendelian Inheritance in Man (OMIM) websites, based at the United States National Library of Medicine and Johns Hopkins University, respectively, for their permission to reproduce screenshots.

# Section One

## General principles of medical genetics

# General principles of medical genetics

## Introduction

This section summarises the basic principles of normal inheritance and genetic disease and outlines the approaches for identification of people with or without risk factors for a genetic disorder.

## Normal human inheritance

Medical genetics is concerned with human biological variation as it relates to health and disease. This variation may be attributable to inherited genetic information (nature) or to environmental factors (nurture). It can also result from combinations of these two influences. The genetic information is coded in DNA, which is packaged into chromosomes. Each chromosome contains a single DNA molecule consisting of two strands woven together as a double helix. Each nucleus has 46 chromosomes; these can be arranged into a karyotype of matching pairs, starting with the largest (numbered 1) down to the smallest (numbered 22) (Figure 1.1). This leaves the sex chromosomes, which are two X chromosomes in a female (Figure 1.1) and an X and a Y in a male (Figure 1.2). When an individual reproduces, only one of each pair will be transmitted to the egg or the sperm. Thus, each egg has only 23 chromosomes (1–22 and an X). Each sperm similarly has 23 chromosomes with one of each pair, 1–22 and either the X or the Y chromosome. Fusion of the egg and sperm restores the full complement of 46 chromosomes and establishes the sex of the embryo.

DNA is composed of four types of bases: adenine (A), cytosine (C), guanine (G) and thymine (T). These bases show specific pairing between the DNA strands of the double helix. A pairs with T and G pairs with C. The unit of length of DNA is a base pair (bp) of AT or CG. One thousand bps is a kilobase (kb) and one million bps is a megabase (Mb). The chromosomes vary in size and contain different amounts of DNA, from 249 Mb in each copy of chromosome 1, to 28 Mb in each copy of chromosome 21.

Information is stored in DNA using the sequence of bases (A, T, C or G) along a DNA strand. These bases are read three at a time as a

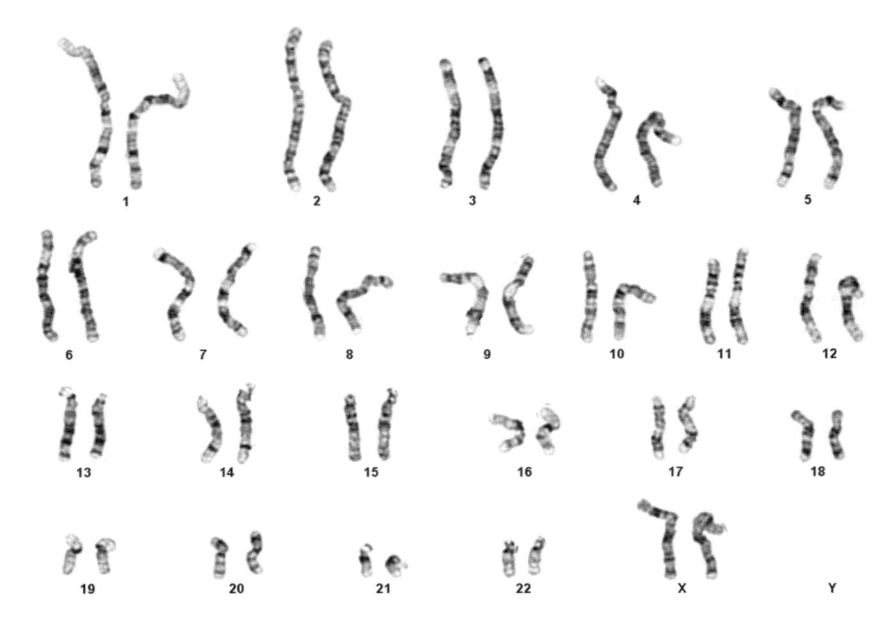


**Figure 1.1** Normal female karyotype

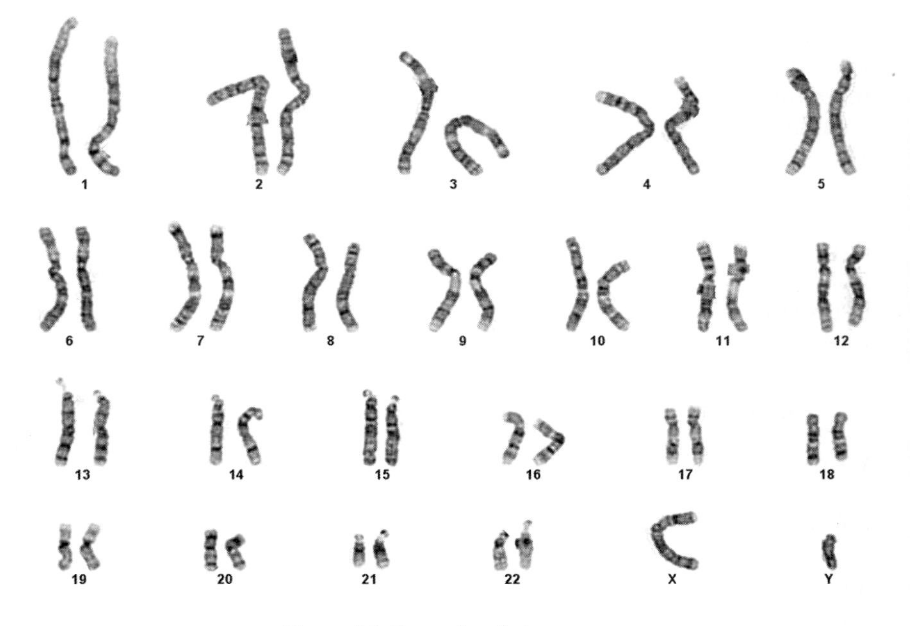


**Figure 1.2** Normal male karyotype

codon and this provides 64 (i.e. $4^3$) combinations. This is more than enough to code for the 20 amino acids that are used to make all proteins and to code for stop and start signals for protein synthesis. The region of the DNA that codes for a particular protein, together with its closely associated regulatory sequences (promoters) and intervening linking sequences (introns), is defined as its gene. Genes vary greatly in size; small genes may be under 1 kb in size, while enormous genes may be over 1 Mb in size. Some areas of the DNA between genes ('enhancers' and 'silencers') have roles in long-range regulation of gene function, but large areas of human DNA have no known function.

In total, approximately 30 000 genes are encoded in human DNA. Most of these have now been identified and their DNA sequence determined, at least in part. Some genes are unique (with a single copy in each chromosome) whereas others are repeated, with multiple copies that may be adjacent or scattered. The genes are not evenly distributed throughout the chromosomes. The dark banded areas of the chromosomes in Figures 1.1 and 1.2 are relatively 'gene poor' compared with the lighter banded areas. More genes are found towards the ends of the chromosomes (telomeres) than around the central constrictions (centromeres).

Not all of the genes encode proteins. In fact, in addition to over 21 500 protein-encoding genes, there are at least 8475 genes that encode RNA molecules. The latter include ribosomal and transfer RNA molecules as well as over 1000 of the more recently recognised regulatory 'micro RNA' molecules that can regulate the function of protein-encoding genes. In contrast to early assumptions, it is now also clear that a single gene can encode more than one protein molecule. This can result from 'alternative splicing' in which the actual protein-encoding segments ('exons') within a gene can be joined together in different arrangements during the process of 'splicing'. During this process, the intervening, non-protein encoding, segments ('introns') are removed from the messenger RNA, prior to the protein synthesis.

## Types of genetic disease

With the exception of identical twins, individuals vary. This variation reflects inherited genetic factors, environmental influences and their interaction. In medical genetics, there are often difficulties in defining the boundary between 'normal' genetic variation and 'mild' genetic disease. It can also be hard to disentangle the influences of one gene on another and of environmental factors on genes. The subdivision into various types of genetic disease is therefore somewhat artificial and the frequency of each group of genetic diseases depends on where the boundaries between normality and disease are placed. Traditionally, genetic diseases are subdivided into:

- chromosomal disorders
- single-gene disorders
- multifactorial disorders
- somatic cell genetic disorders.

## Chromosomal disorders

By definition, a chromosomal disorder is present if there is a visible alteration in the number or structure of the chromosomes. These changes may affect either the sex chromosomes (X or Y) or the autosomes (numbers 1–22). Using routine light microscopy, multiple newborn cytogenetic surveys have revealed a frequency of six chromosomal disorders per 1000 births. Of these, approximately two-thirds result in either mental or physical disability. The liveborn infants with chromosomal abnormalities represent only a small proportion of all chromosomally abnormal conceptions. It is estimated that the rate of chromosomal abnormality in embryonic and fetal deaths is within the range 32–42%; the proportion of all recognised conceptuses that are chromosomally abnormal is 5–7%.

Normally, each sperm and egg has 23 chromosomes, with one of each pair of autosomes (1–22) and one sex chromosome. Malsegregation is common and can result in an egg or a sperm with either an extra copy of a chromosome (24 in total) or a missing chromosome (22 in total). In general, loss of chromosomal material is more serious than possession of additional material and autosomal imbalances are more serious than sex chromosome imbalances. In consequence, the types of chromosomal abnormality that predominate in miscarriages and in liveborn infants are different.

Additional copies of any autosome in the egg or sperm will result in 47 chromosomes in the fetus and these will generally result in miscarriage. For example, an extra copy of chromosome 16 (trisomy 16) is the most common autosomal trisomy in miscarriages, whereas trisomies for chromosome 21 (Down syndrome, Figure 1.3), 18 (Edwards syndrome, Figure 1.4) and 13 (Patau syndrome, Figure 1.5) are the most common trisomies in liveborn infants.

A missing autosome usually results in a very early pregnancy failure and is undetected. However, monosomy X with a single copy of the X chromosome and 45 chromosomes in total (45,X, Figure 1.6) occurs

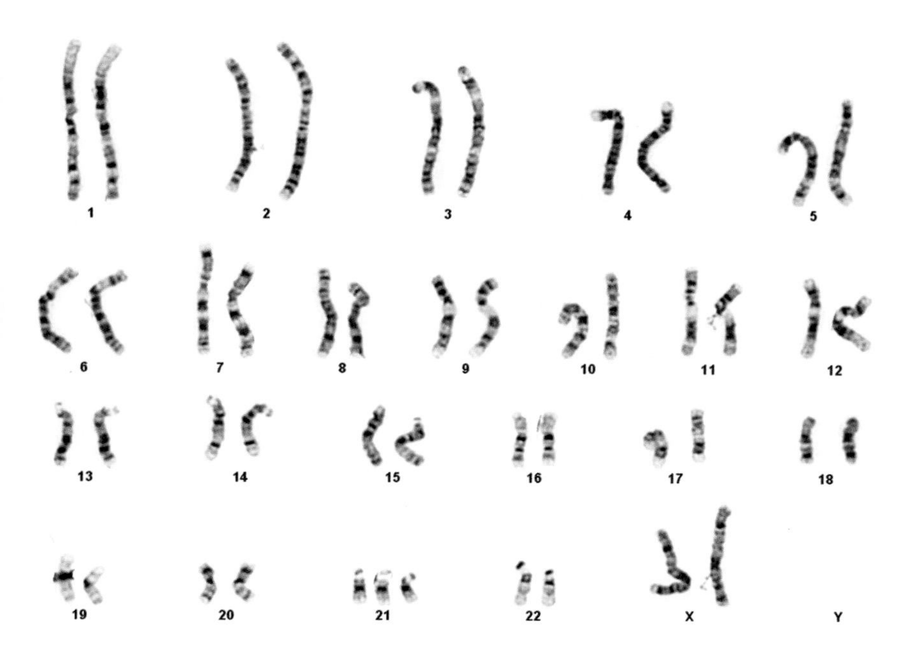


**Figure 1.3** Trisomy 21 female

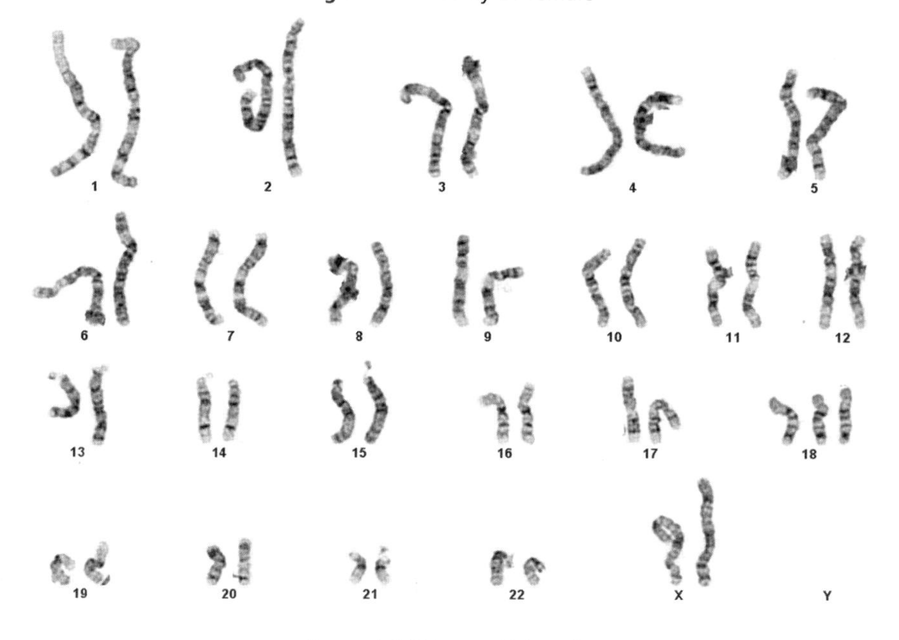


**Figure 1.4** Trisomy 18 female

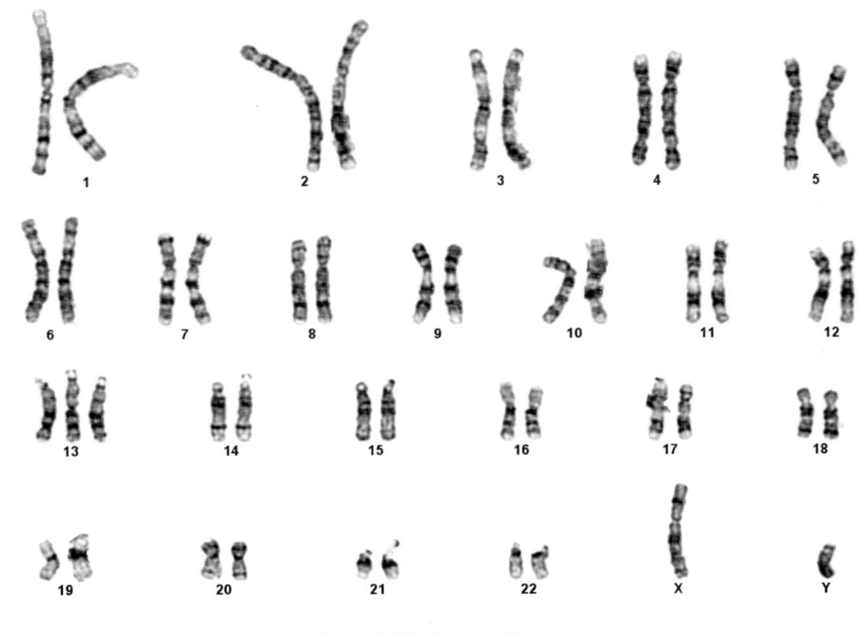


**Figure 1.5** Trisomy 13 male

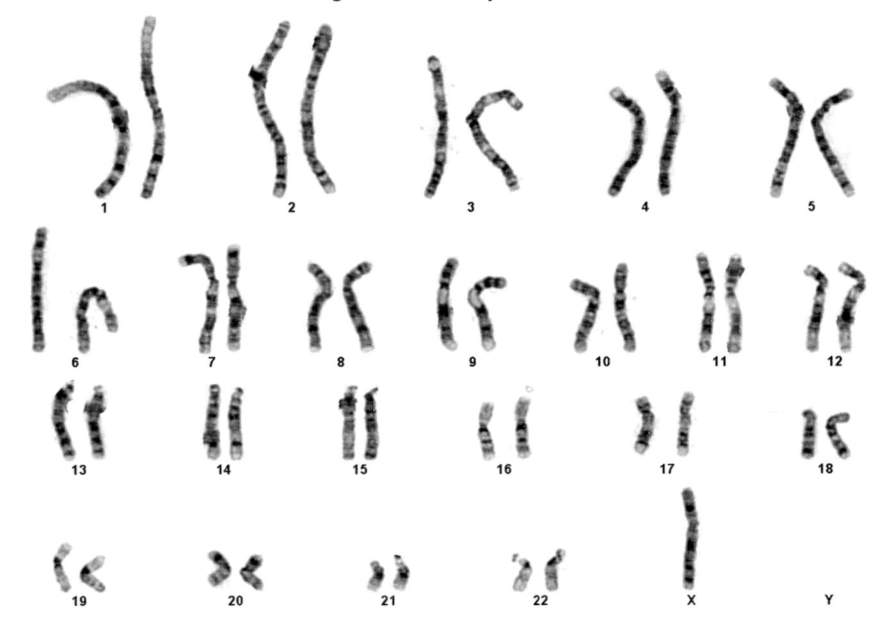


**Figure 1.6** Turner syndrome: 45,X

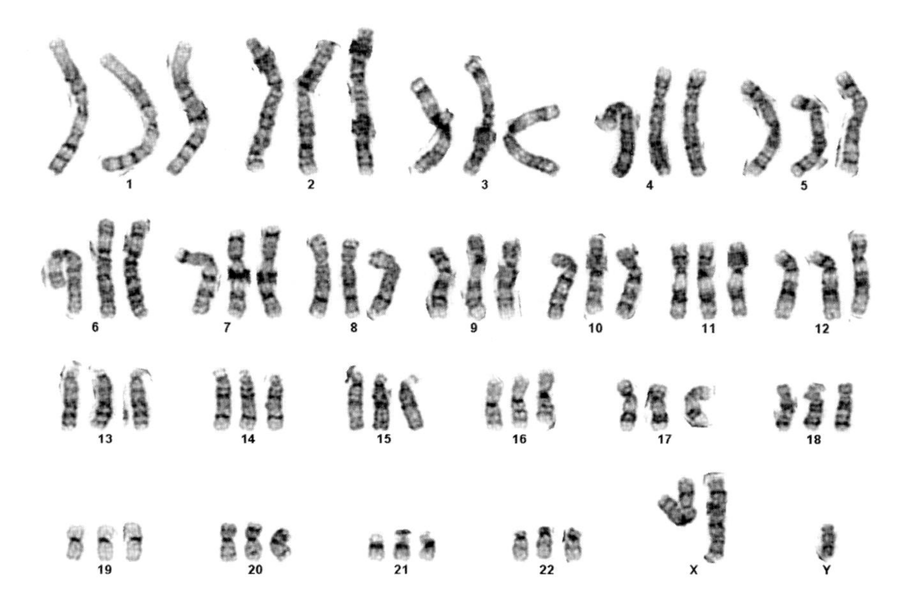


**Figure 1.7** Triploidy: 69,XXY

in approximately 1% of all conceptions although it is believed that over 98% of these pregnancies are spontaneously miscarried. This high lethality in the womb contrasts with the relatively mild postnatal features of children with 45,X (Turner syndrome).

An extra half set of chromosomes results from fertilisation of an egg by two sperm and leads to 69 chromosomes in total (triploidy, Figure 1.7). Triploidy usually results in a miscarriage but, exceptionally, infants may be liveborn.

Structural aberrations result from chromosomal breakage. When a chromosome breaks, two unstable sticky ends are produced. Generally, repair mechanisms rejoin these two ends. However, if more than one break has occurred, there is the possibility of joining the wrong ends as the repair mechanisms cannot distinguish one sticky end from another. The most common types of structural aberrations are translocations and deletions.

Translocations involve the transfer of chromosomal material between chromosomes. The process requires simultaneous breakage of two chromosomes, which then repair in an abnormal arrangement (Figure 1.8). This exchange can involve any two chromosomes and usually results in no loss of vital DNA. In this case, the individual is usually clinically normal

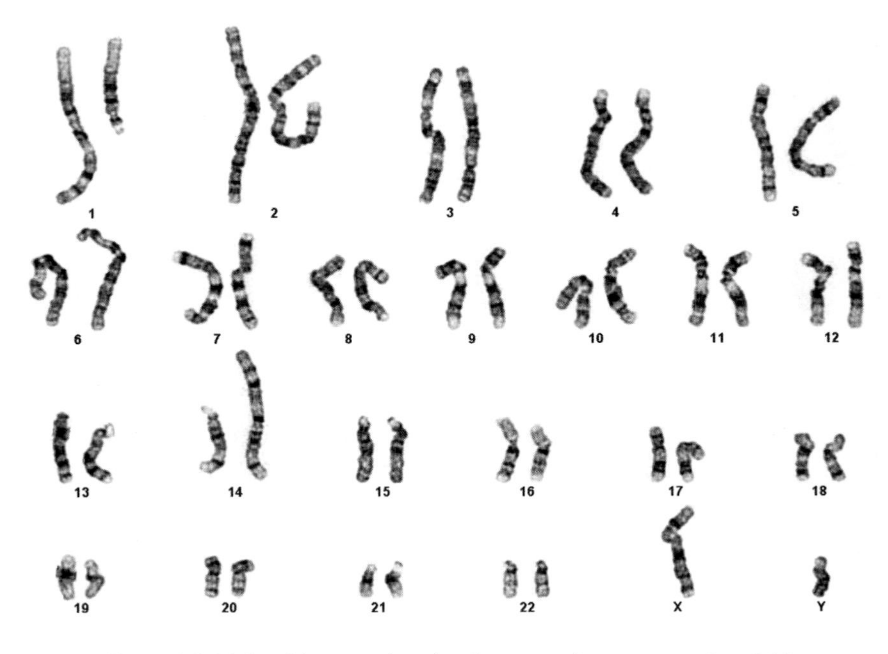


**Figure 1.8** Male with a translocation between chromosomes 1 and 14

and is said to have a balanced translocation. The medical significance is, then, its relevance for future generations, because the carrier of a balanced translocation is at risk of producing chromosomally unbalanced offspring. The overall frequency of translocations in the general population is two per 1000.

Deletions arise from loss of chromosomal material between two breakpoints on the same chromosome. They can also result from a parent with a balanced translocation. Deletions of parts of the autosomes almost always produce clinical effects if they are visible by light microscopy. These clinical effects commonly include learning disabilities, congenital malformations and unusual facial features. Submicroscopic deletions (i.e. those that are too small to be detected by light microscopy) are also clinically important. The smallest visible change to a chromosome using the light microscope approximates to the loss of 4 Mb of DNA. Smaller losses (submicroscopic deletions or 'microdeletions') can easily include multiple genes and a large number of microdeletion syndromes have now been identified using DNA probes for the deleted area (Table 1.1). Geneticists may organise tests for such microdeletions (or microduplications)

**Table 1.1 Examples of chromosomal microdeletion syndromes**

| *Disorder* | *Region of microdeletion** |
|---|---|
| Alagille syndrome | 20p12 |
| Angelman syndrome | 15q11–12 |
| DiGeorge/VCFS syndrome | 22q11 |
| Miller-Dieker lissencephaly | 17p13 |
| Prader–Willi syndrome | 15q11–12 |
| Rubinstein-Taybi syndrome | 16p13 |
| Smith-Magenis syndrome | 17p11 |
| Williams syndrome | 7q11 |
| Wilms tumour, aniridia syndrome | 11p13 |

* p represents the short arm of a chromosome, which is above the centromere in the karyotype; q represents the long arm of a chromosome, which is below the centromere in the karyotype. VCFS = velocardiofacial syndrome.

using techniques such as fluorescence *in situ* hybridisation (FISH), multiple ligation-dependent probe amplification (MLPA) and/or aCGH (see below in the sections entitled DNA analysis and Chromosome analysis).

## Single-gene disorders

Single-gene disorders (Mendelian disorders) are the result of mutations in one or both members of a pair of autosomal genes or mutations in single genes on the X or Y chromosomes. Within each chromosome the genes have a strict order, with each gene occupying a specific location or locus. Thus, the autosomal genes are present in pairs, one on the maternally inherited copy of the pair and the other on the paternal copy. Alternative forms of a gene are called alleles and these arise by mutation of the normal allele and may or may not have an altered function. If both members of a gene pair are identical then the individual is said to be homozygous (noun – homozygote) for that locus and if different the individual is said to be heterozygous (noun – heterozygote).

Any gene-determined characteristic is called a trait. If a trait is manifest in the heterozygote then the trait is described as dominant, whereas if it only occurs in the homozygote then it is described as recessive. Hence, single-gene disorders may be classified according to their chromosomal location (autosomal, X-linked or Y-linked) and further subdivided into dominant and recessive (Table 1.2).

**Table 1.2 Examples of single-gene disorders (in approximately decreasing frequency order within each type of mechanism of inheritance)**

| *Mechanism of inheritance* | *Single-gene disorders* |
|---|---|
| Autosomal dominant | Inherited breast cancer, inherited colon cancer, dominant otosclerosis, familial hypercholesterolaemia, von Willebrand's disease, adult polycystic kidney disease, neurofibromatosis, myotonic dystrophy, Huntington disease, tuberous sclerosis |
| Autosomal recessive | Cystic fibrosis, recessive learning disabilities (multiple subtypes), congenital deafness (multiple subtypes), phenylketonuria, spinal muscular atrophy |
| X-linked dominant | Xg blood group, vitamin D-resistant rickets, hereditary motor and sensory neuropathy (one type), incontinentia pigmenti, Rett syndrome |
| X-linked recessive | Red–green colour blindness, nonspecific X-linked learning disabilities (several subtypes), fragile X syndrome, Duchenne muscular dystrophy, haemophilia A, X-linked ichthyosis |
| Y-linked | Swyer syndrome (XY gonadal dysgenesis) |

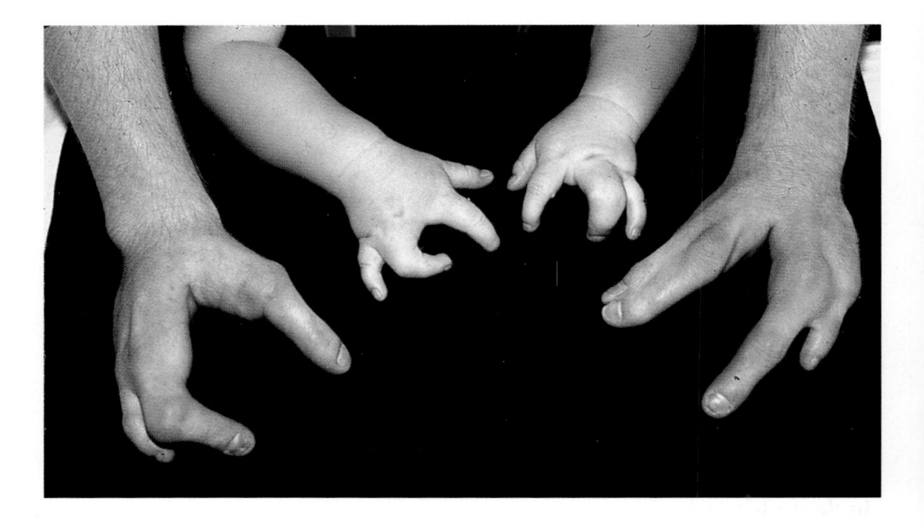

**Figure 1.9** Father and daughter with split-hand syndrome (ectrodactyly)

## AUTOSOMAL DOMINANT INHERITANCE

Autosomal dominant inheritance is illustrated by the condition shown in Figure 1.9. In this family, the father and daughter have deformed hands and feet with longitudinal splitting, which is called ectrodactyly. This condition is inherited as an autosomal dominant trait and, in this family,

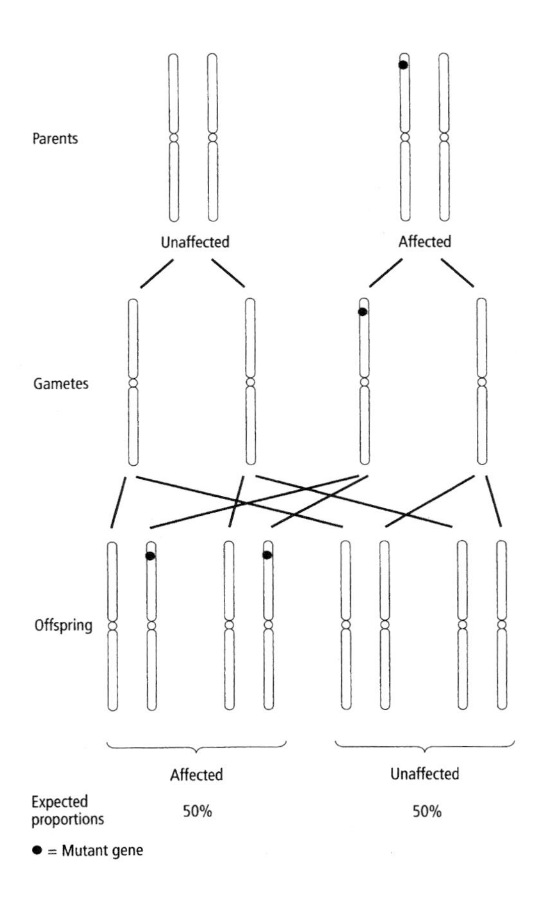


**Figure 1.10** Diagram of autosomal dominant inheritance

it is caused by a mutation in one copy of the paired autosomal genes called *TP63* (tumour protein p63). This gene (previously known as *TP73L*) is located on the long arm of chromosome 3. Each affected person has a single underactive copy and is affected despite the presence of the normal copy of this gene on the opposite copy of chromosome 3, which was inherited from the other parent. The risk of an affected person transmitting the copy of chromosome 3 with the mutated gene is one in two (Figure 1.10), each time that he or she has a child.

Variable clinical severity in different affected family members is common in autosomal conditions and is termed 'variable expressivity'. In addition, in several autosomal dominant conditions, individuals can possess the mutated gene copy but be clinically completely unaffected (which is known as 'incomplete penetrance').

## AUTOSOMAL RECESSIVE INHERITANCE

Autosomal recessive inheritance is illustrated by the condition shown in Figure 1.11. In this family, the parents and other relatives are healthy but one of their children has learning disabilities attributable to Hurler syndrome. Hurler syndrome is inherited as an autosomal recessive trait and is caused by a mutation in both copies of the alpha-L-iduronidase gene (symbolised *IDUA*) located on the tip of the short arm of chromosome 4. The affected child has two copies of the mutant gene whereas the normal parents (who each have a single mutant copy and a normal gene on the opposite copy of chromosome 4) are termed carriers or heterozygotes. For two carrier parents the chance of having a further affected child is one in four (i.e. 25%) each time that they have a child together (Figure 1.12).

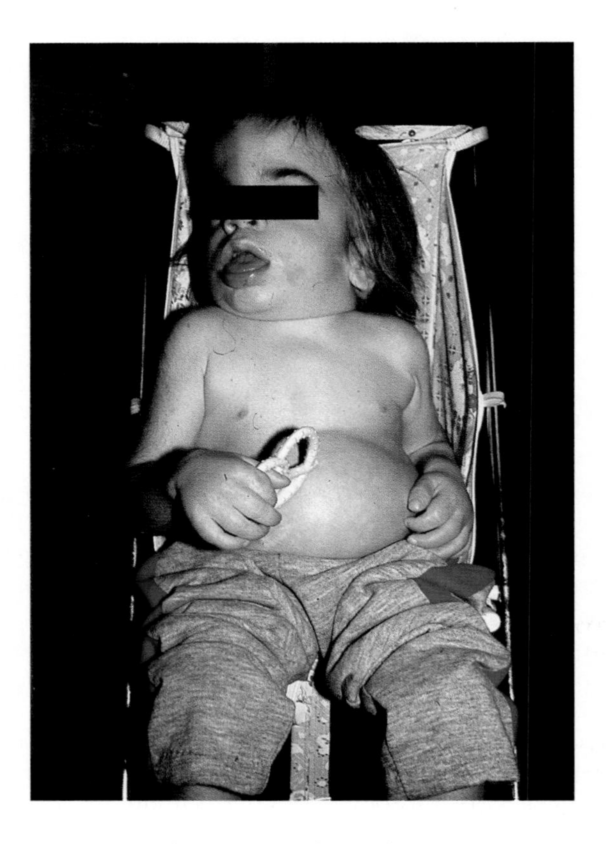

**Figure 1.11** Hurler syndrome

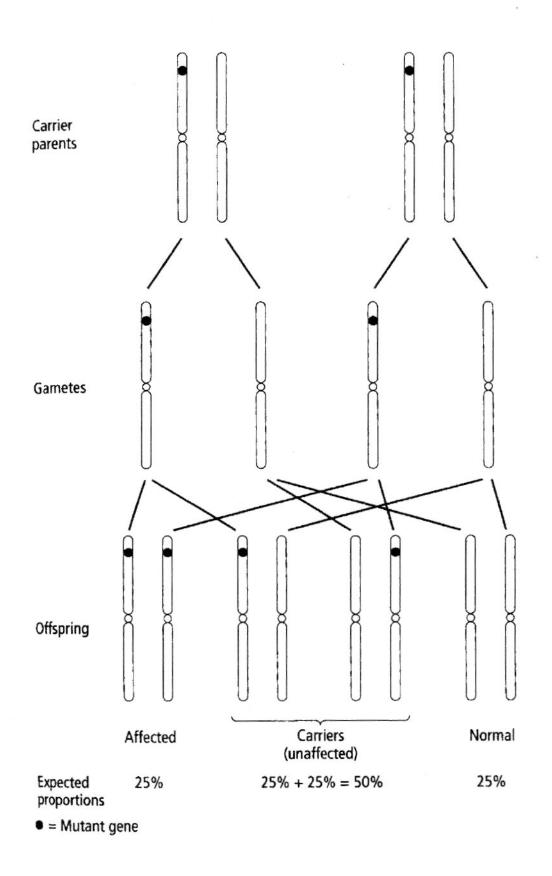


**Figure 1.12** Diagram of autosomal recessive inheritance

## X-LINKED RECESSIVE INHERITANCE

X-linked recessive inheritance is illustrated by the condition shown in Figure 1.13. The boy shown in the figure has learning disabilities caused by Lesch–Nyhan syndrome. This is caused by a mutation in the *HPRT* gene on the long arm of the X chromosome. Males with a single mutant *HPRT* gene are affected, as the partner sex chromosome (Y) does not contain a copy of this gene. Females with a single mutant *HPRT* gene are healthy owing to the presence of the normal *HPRT* gene on their other X chromosome. Such a female, with one normal and one mutant copy, is termed a carrier and she has a one in two (i.e. 50%) chance of transmitting the X chromosome incorporating the mutant gene to the next generation each time that she has a child. The sex of each child is

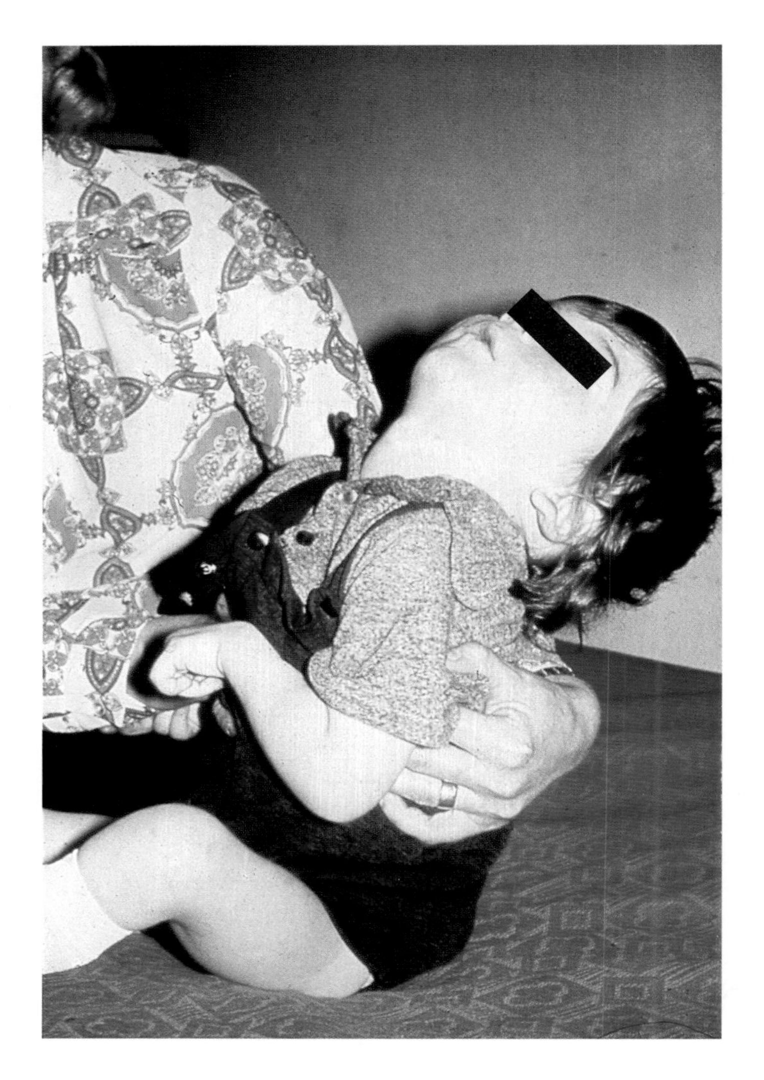

**Figure 1.13** Lesch–Nyhan syndrome

determined by the sex chromosome transmitted by the father. There is a one in two chance that the Y chromosome will be transmitted (resulting in a male) and a one in two chance that the X chromosome will be transmitted (resulting in a female). Thus, for a carrier mother half of her sons will be affected and half of her daughters will be carriers (Figure 1.14).

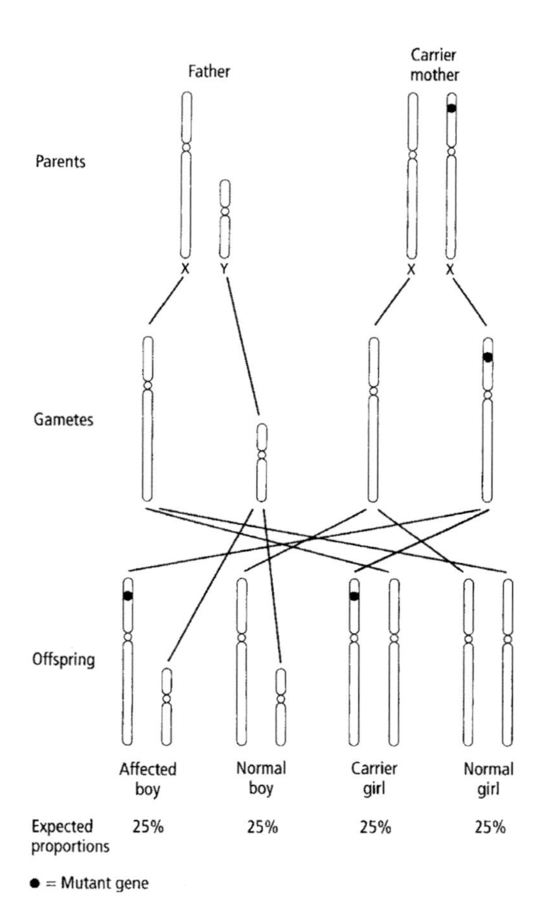


**Figure 1.14** Diagram of X-linked recessive inheritance

The overall frequency of single-gene diseases is not known. An early estimate, which is now known to be much too low, gave a combined frequency of ten per 1000 (seven per 1000 autosomal dominant, 2.5 per 1000 autosomal recessive, plus 0.5 per 1000 X-linked recessive). The number of recognised single-gene traits has now increased enormously to over 20 000 and includes several common conditions such as familial breast cancer (which affects two to six per 1000, depending on ethnicity). In addition, DNA analysis has revealed higher than expected frequencies of generally asymptomatic people with one or two mutant alleles at a disease-associated gene locus. For example, up to 1% of the population has a mutant allele for von Willebrand's coagulation factor, and around one in 400 possess two mutant haemochromatosis alleles, yet most of these individuals are asymptomatic.

## Multifactorial (or part-genetic) disorders

Multifactorial disorders result from an interaction of one or more genes with one or more environmental factors. In effect, the genetic contribution predisposes the individual to the actions of environmental agents. Such an interaction is suspected when conditions show an increased recurrence risk within families, which does not reach the level of risk seen for single-gene disorders. In contrast to single-gene disorders, the pedigree pattern is not diagnostic for multifactorial disorders and evidence for this pattern of inheritance may be derived from twin studies.

Identical (or monozygotic) twins have 100% of their genes in common whereas nonidentical (dizygotic) twins have only 50% of their genes in common. If a condition has no genetic influence then the concordance frequencies (i.e. the observed frequency of the condition being present or absent in both members of the pair of individuals) in identical and in nonidentical twins should be similar. For multifactorial disorders, the concordance frequency in identical twins is higher than in nonidentical twins, but is not as high as the 100% figure seen with single-gene disorders. Thus, for example, it is observed that the likelihood that both nonidentical twins will be affected with multifactorial cleft lip and palate is 5%, whereas for identical twins the comparable figure is 35%.

Multifactorial disorders are believed to account for approximately 50% of all congenital malformations and are thought to be relevant to many common chronic disorders of adulthood, including hypertension, rheumatoid arthritis, schizophrenia, manic depression, multiple sclerosis, diabetes mellitus, premature vascular disease and senile dementia. The collective frequency of multifactorial malformations is estimated to be 45 per 1000 live births and the collective frequency of multifactorial common disorders of adulthood is probably at least 600 per 1000 over a lifetime. In addition, multifactorial inheritance is suspected for many common psychological disorders of childhood, including dyslexia, specific language impairment and attention deficit hyperactivity disorder. Hence, multifactorial disease represents the most common type of genetic disease in both children and adults.

The genetic component of multifactorial disorders is usually composed of one or more DNA sequence variants that are individually of small effect. These so-called single nucleotide polymorphisms (SNPs) are, however, generally much more prevalent in the population than the mutations associated with single-gene disorders.

## Somatic cell (or cumulative) genetic disorders

When a mutation is present in the fertilised egg it will be transmitted to all daughter cells, including the germ cells. If, however, a mutation arises after the first cell division, this mutation will only be found in a proportion of cells and the individual is said to be mosaic. The mutation may be confined to some of the gonadal cells (gonadal mosaic) or to the somatic cells (somatic mosaic) or may occur in a proportion of both.

Cancers are usually somatic cell genetic disorders. The initiating event for each sporadic (nonfamilial) cancer is the occurrence of one or more key mutations in the same somatic cell. With progression, further genetic changes accumulate in the cancerous cells and can include mutations in other genes as well as numerical and structural chromosomal abnormalities (Figure 1.15). These changes are confined to the somatic cells of the tumour and not the individual's gonadal cells. Thus the condition is not inherited and the risks to family members are not increased above the general population risk by these somatic cell genetic changes.

For some rare cancers (such as retinoblastoma) and in some families with common cancers, the first step in the cascade of mutations may be

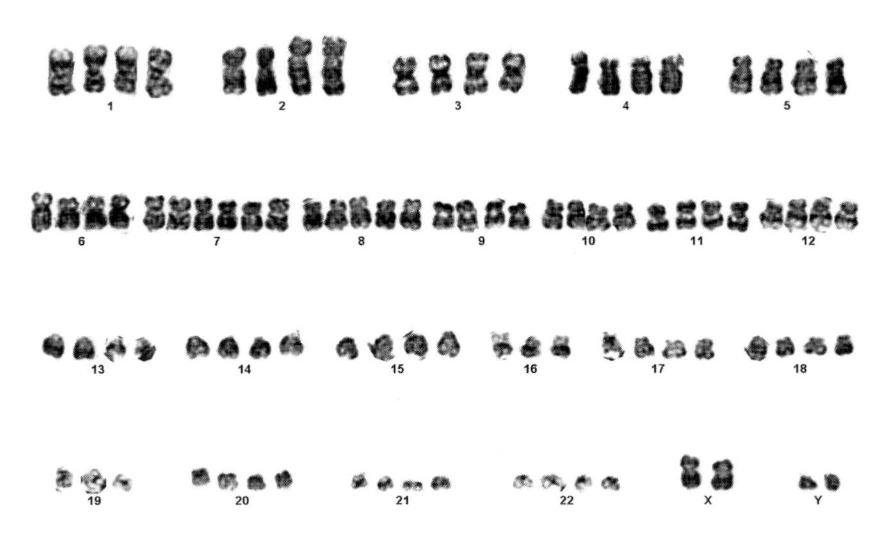


**Figure 1.15** Multiple numerical and structural chromosomal aberrations in a rhabdomyosarcoma

inherited. In this situation, the condition is inherited as a single-gene disorder, with high risks to other family members.

Noninherited cancers are common, with a lifetime risk of one in three. Somatic cell genetic disorders might also be involved in other clinical conditions such as autoimmune disorders and the ageing process.

## Drawing the family tree

Normally, a family will see a genetic specialist after being referred by a general practitioner, another specialist or a member of one of the professions allied to medicine. The referring healthcare professional may have the genetic basis of a condition brought to his or her attention by a patient or may find a clinical sign suggestive of a genetic disorder on examination. Once alerted to the possibility of a genetic disorder, even the nongeneticist should draw a family tree. This is performed freehand, using a standard set of symbols (Figure 1.16).

The family tree, or pedigree, is a compact way of storing a large amount of family information. Each generation occupies the same horizontal level and within a generation the birth order is presented from left to right. It is usually easiest to start with the youngest generation at the bottom of the page and then work back to older generations. For each member of the pedigree, name and age are usually included. Miscarriages, neonatal deaths, children with physical or learning disabilities and parental consanguinity might not be mentioned by the patient unless enquiries are specifically made to elicit such information.

The newborn child shown in Figure 1.17 has bilateral polydactyly. The family history taken at admission for delivery is shown in Figure 1.18. The family tree was then taken properly and this revealed multiple affected individuals in several generations, including the child's mother (Figures 1.19 and 1.20).

## Interpreting the family tree

In the family with polydactyly shown in Figure 1.20, both males and females are affected and so the gene causing polydactyly is highly unlikely to be on the Y chromosome. There is also an instance of a male passing it on to a male. Since a male does not transmit an X chromosome to his sons, we can conclude that the polydactyly gene is not on the X chromosome and therefore it must be on one of the autosomes.

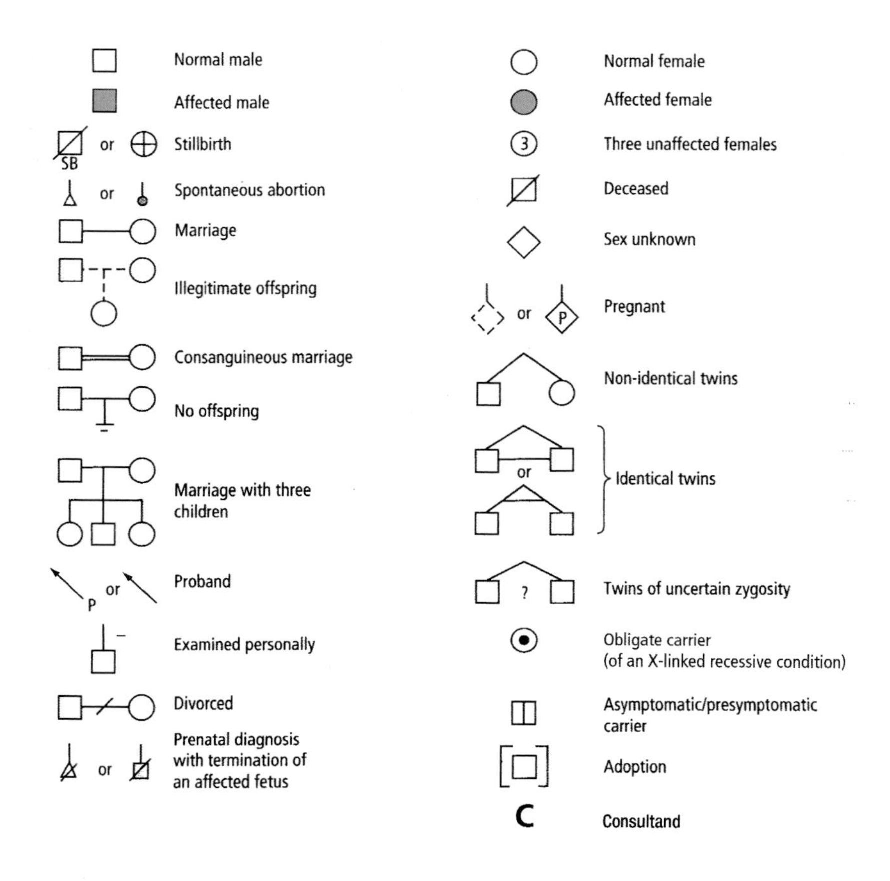


**Figure 1.16** Symbols used for pedigree construction

Looking at the family, it would be extremely unlikely that the partners of different affected people were, by chance, carriers of the polydactyly gene alteration and so affected children must have received only a single copy of the abnormal gene from the affected parent. Since they have developed polydactyly even though they have inherited a healthy gene from the other parent, the polydactyly copy of the gene is said to be dominant to the healthy copy of the gene. Polydactyly in this family is thus inherited as an autosomal dominant trait.

Each affected person with this or any other autosomal dominant trait is thus generally heterozygous, with one normal and one mutant gene at this genetic location. The affected person will pass on either the normal or the mutant gene and so the risk to each child is one in two (Figure 1.10, page 13).

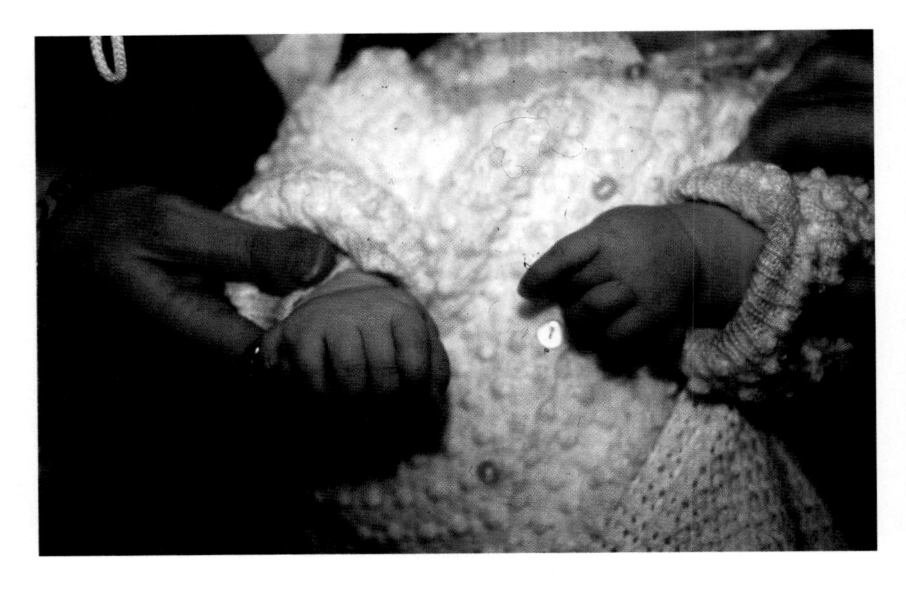

**Figure 1.17** Bilateral postaxial (little finger side) polydactyly

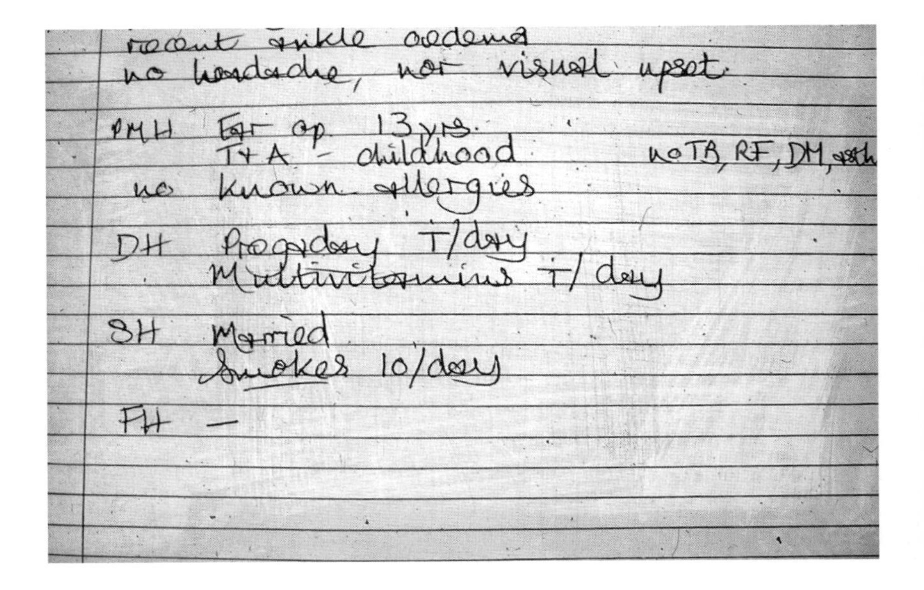

recent ankle oedema
no headache, nor visual upset.

PMH Ear op. 13yrs.
T+A – childhood    no TB, RF, DM, asth
no known allergies

DH Pregaday 1/day
Multivitamins 1/day

SH Married
Smokes 10/day

FH —

**Figure 1.18** How not to take a family history

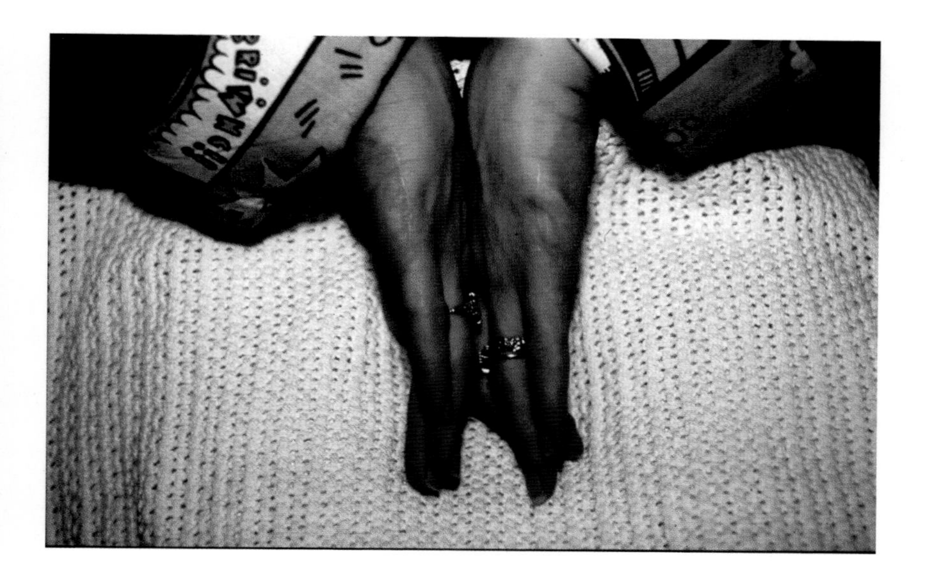

**Figure 1.19** Mother of the child in Figure 1.17, with scars where her extra digits were removed

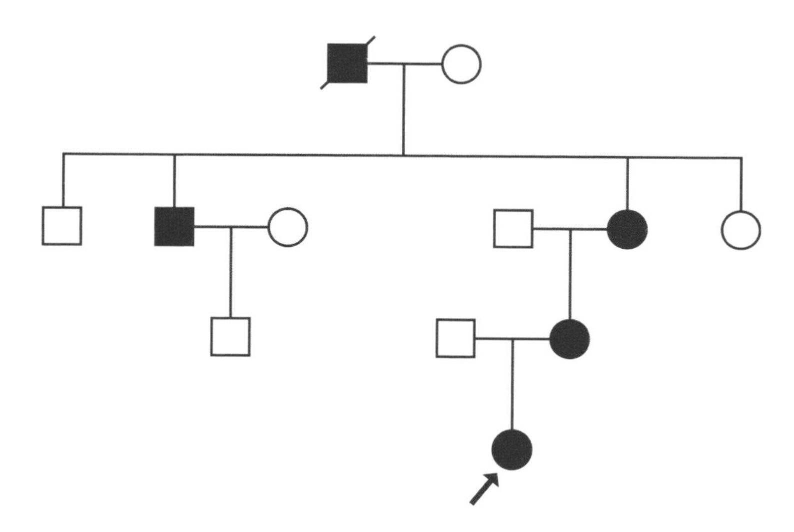

**Figure 1.20** Pedigree of mother and child in Figures 1.17 and 1.19

In a family tree, the following features help to confirm the autosomal dominant mode of inheritance:

- There are usually affected people in each generation with passage from one affected person to another (this is termed a vertical pattern of inheritance).
- Men and women are equally likely to be affected and, if affected, have a similar severity.
- Men can transmit the condition to sons or daughters.
- Women can transmit the condition to sons or daughters.
- On average one-half of the children of an affected person will be affected.

Figure 1.21 shows a different pattern: autosomal recessive inheritance. In contrast to the previous example, affected people are present in only one sibship – that is, a family group of brothers and sisters – in one generation. This is called a horizontal pattern of inheritance. The parents are unaffected but may be blood relatives (consanguineous). Both men and women may be affected, and with a similar severity, but with a low risk to their own offspring.

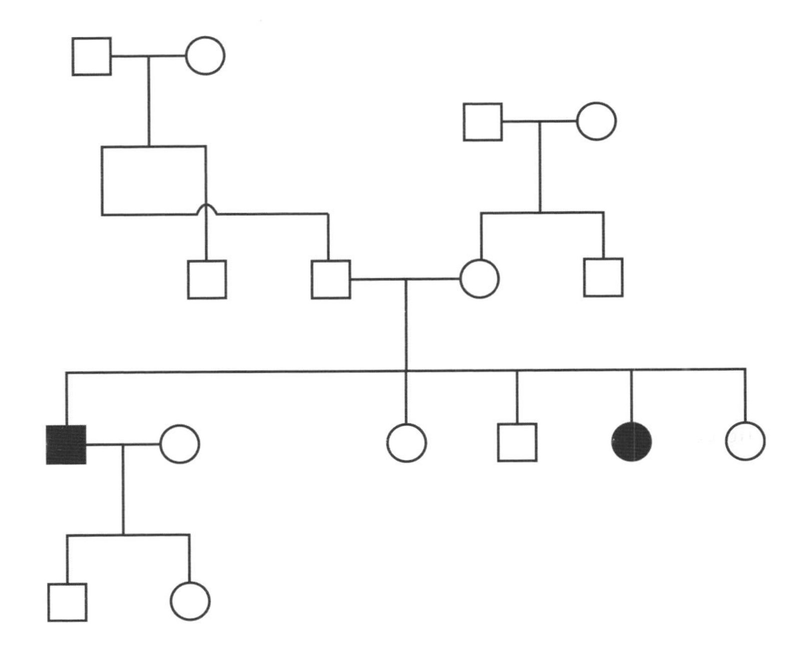

**Figure 1.21** Example of an autosomal recessive family tree

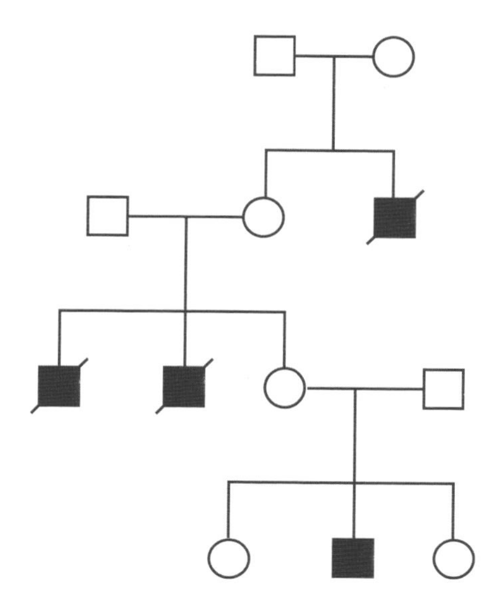

**Figure 1.22** Example of an X-linked recessive family tree

Another pattern of inheritance can be seen in Figure 1.22. In contrast to the autosomal dominant and autosomal recessive patterns, only males are affected. Affected men occur in more than one generation and are linked to each other by healthy (or mildly affected) women (this is termed a 'knight's move' pattern of inheritance). Affected men never transmit the trait to their sons (that is, there is no male-to-male transmission). The women who link affected males are termed obligate carriers, as they must be heterozygous for the mutant gene.

## DNA analysis

DNA can be extracted from any nucleated tissue sample, including blood leucocytes, buccal mucosal cells, skin fibroblasts, amniotic fluid cells and chorionic villus samples. A 10 ml blood sample anticoagulated with ethylenediamine tetra-acetic acid (EDTA) should be maintained at ambient temperature. Once extracted, the DNA is stored frozen, in which state it is stable and available for future analysis for many years. This can be particularly important for lethal inherited conditions as, for some, a test is not yet available or a test might only be required in the future, long after the affected person(s) in the family has died. DNA can be extracted from stored pathological material (e.g. paraffin blocks), but the quality is not as good as from a fresh sample.

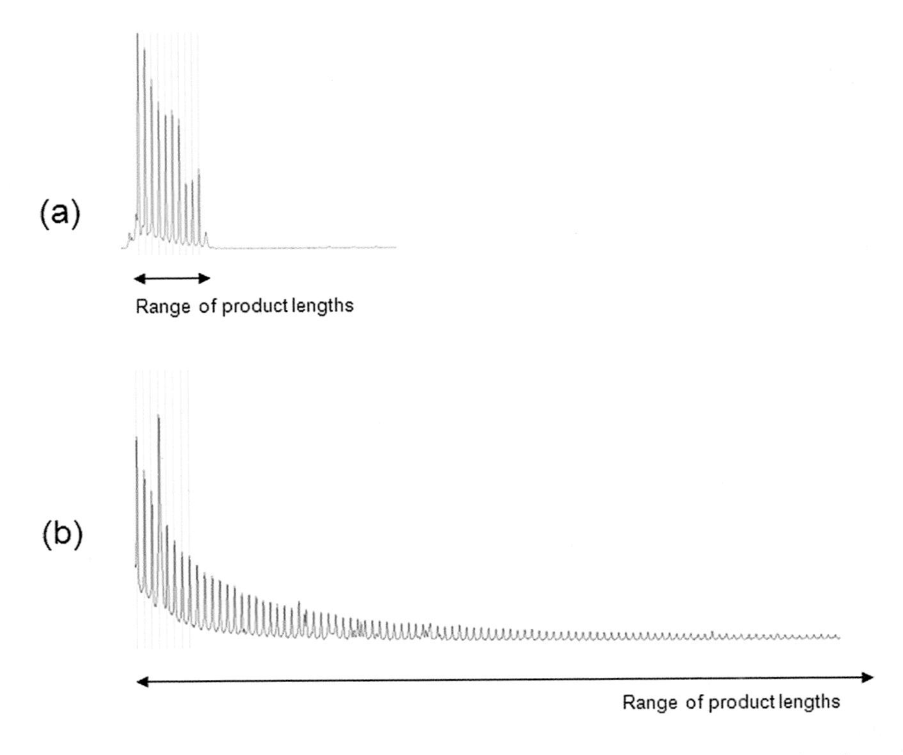


**Figure 1.23** DNA analysis for myotonic dystrophy type 1, by triplet repeat-primed polymerase chain reaction, to detect an enlarged and thus pathogenic triplet repeat tract. A polymerase chain reaction (PCR) primer that binds to the triplet repeat itself is used. The analysis of the unaffected individual's DNA is shown in panel a. A much wider range of PCR product sizes results from the affected patient (panel b) owing to the enlarged triplet repeat tract. Adapted from images kindly provided by Dr Alexander Cooke, Glasgow.

Mutations of DNA are divided into length mutations (with gain or loss of DNA) and point mutations (which alter the genetic code at a single base).

In all point mutations, and most length mutations, the mutation is stable and each affected person in the family will have an identical mutation. In some disorders caused by length mutations, the mutation is unstable and varies in size between family members. Figure 1.23 shows an example of DNA analysis by a highly specialised technique, triplet repeat-primed PCR (TP-PCR), for an unstable length mutation that consists of many CTG triplet repeats in the gene for myotonic dystrophy (an autosomal dominant form of muscular dystrophy). TP-PCR is an unusual form of PCR in that one of the primers is designed to bind within the repetitive sequence,

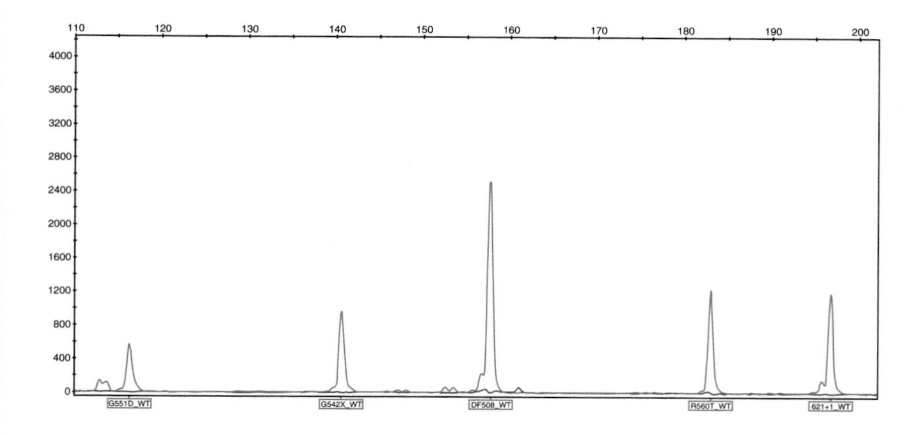


**Figure 1.24** Fluorescent amplification refractory mutation system DNA test for the detection of any of five possible cystic fibrosis gene (*CFTR*) mutations in neonates with raised immunoreactive trypsinogen levels. The homozygous normal results at each position are shown here as green peaks.

with just one primer binding to a specific unique sequence position. The PCR thus results in a range of product sizes, owing to the slightly different binding locations of the repeat sequence primer. Pathogenically large repeat tracts result in an abnormally large range of PCR product sizes (panel b). The products can be visualised on an automated DNA sequencer, owing to the fluorescent label attached to one of the PCR primers.

Figures 1.24 and 1.25 show an example of the use of the fluorescent amplification refractory mutation system (ARMS) technique of DNA analysis for five specific mutations in neonates who are suspected (from raised immunoreactive trypsinogen [IRT] levels) to have cystic fibrosis. Different PCR primers are designed to bind specifically to normal sequences (giving the green peaks) or to mutations (resulting in the blue peaks). Samples are then run on a DNA sequencer to permit differentiation of the products. In Figure 1.24, the neonate does not possess any of the five *CFTR* gene mutations. In a carrier for cystic fibrosis there may be one peak for the normal gene and one peak for the mutated gene copy. In Figure 1.25, the neonate is compound heterozygous with two different mutant *CFTR* gene copies, G551D and delta F508 (a 3 bp deletion). This results in a blue (mutation) peak superimposed on a green (normal sequence) peak at the G551D position and also blue and green peaks at the delta F508 (or "F508del") position.

Figures 1.26 and 1.27 show an example of a point mutation in the *SDHD* gene, which causes the autosomal dominantly inherited predisposition to phaeochromocytomas. The DNA sequence of the affected region of the

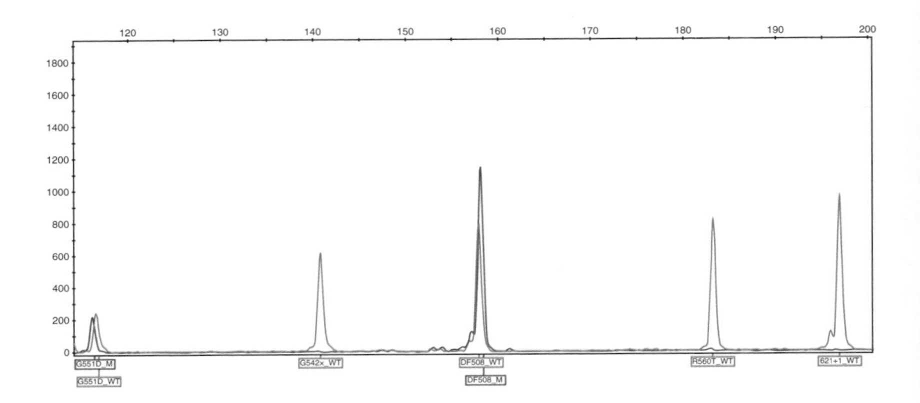


**Figure 1.25** Fluorescent amplification refractory mutation system DNA test for detection of five possible cystic fibrosis gene (*CFTR*) mutations. The neonate whose results are shown here possesses two different CFTR mutations, G551D and delta F508 (F508del). Mutations are shown as blue peaks superimposed on the green peaks, representing products from the mutant and normal gene copies, respectively.

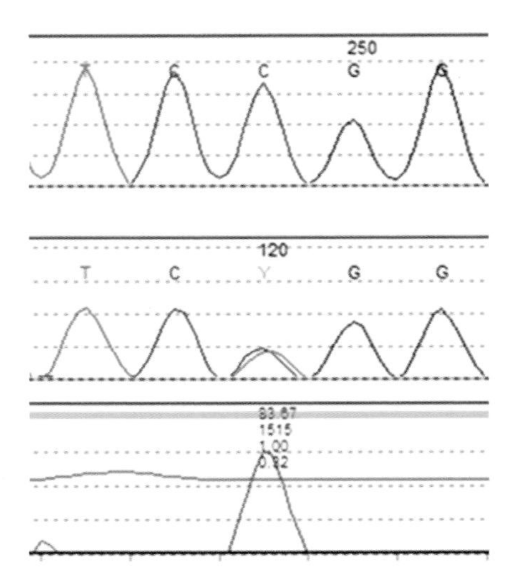


**Figure 1.26** DNA sequencing result (by the Sanger method) for a selected part of the *SDHD* gene's coding region, showing the normal reference sequence (top), the sequence of the patient's blood DNA trace (middle) and a computed prediction of mutation likelihood (bottom). The superimposed red and blue peaks in the middle result from a heterozygous point mutation (changing a C to a T).

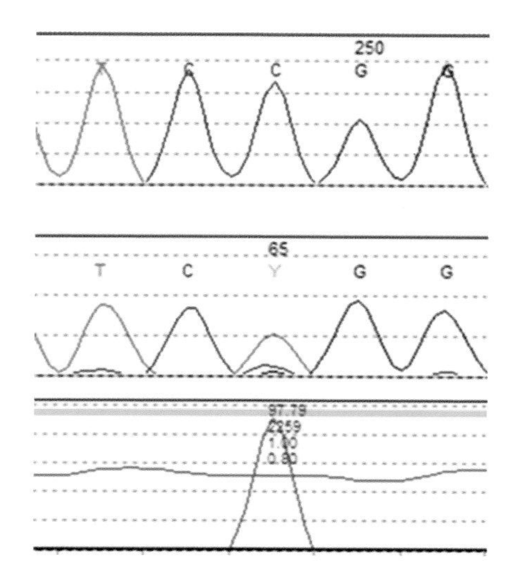


**Figure 1.27** DNA sequencing result for the tumour DNA of the patient whose blood DNA result is shown in Figure 1.26. At the position of the mutation, there is loss of the normal C so that the red peak representing the T is predominant. This 'loss of heterozygosity' is commonly seen in tumour DNA for tumour suppressor genes such as *SDHD*.

gene is shown for the blood DNA (Figure 1.26) and the tumour DNA (Figure 1.27), with coloured peaks representing each base in the DNA sequence, with normal reference sequence in the top section and patient's sample sequence in the middle section. The lower section contains a computer-predicted plot of the likelihood of a mutation being present at each position. The mutation in the affected woman whose blood DNA is analysed in Figure 1.26 is the substitution of a T in place of a C. She also possesses another, normal, copy of the gene as the gene is present on an autosome (i.e. not an X or a Y chromosome) and is therefore heterozygous, with both a C and a superimposed T present in the sequencing result at that position. The 'Y' indicates the presence of either a C or T at this position. The nucleotides on either side appear as just single peaks, as both copies of the gene contain the normal reference sequence nucleotides at those other positions. The tumour DNA result (Figure 1.27), by contrast, shows loss of the normal copy of the gene with, now, predominantly the mutant T at that position. This 'loss of heterozygosity' (or LOH) in the tumour DNA compared with the blood DNA is a commonly observed phenomenon for tumour suppressor genes such as *SDHD*. The loss of the normal copy of the gene in the cells that already contain

**Table 1.3 Indications for DNA analysis**

| *Time of DNA analysis* | *Indications* |
|---|---|
| Prenatally | Diagnosis of at-risk pregnancy (e.g. if a family history of Duchenne muscular dystrophy) Rapid detection of aneuploidy (especially trisomy 21, 18 and 13 or sex chromosome abnormalities XXY or XO) by quantitative fluorescent polymerase chain reaction |
| Neonatally | Confirmation of diagnosis (e.g. cystic fibrosis detected by newborn biochemical screening) and rapid detection of aneuploidy as above (when suspected clinically) |
| Children | Investigation of learning disabilities: exclusion of fragile X syndrome and (particularly when dysmorphic features and/or multiple malformations are also present) detection of submicroscopic DNA duplications and deletions e.g. using array comparative genomic hybridisation<br>Confirmation of diagnosis of a childhood-onset disorder when suspected (e.g. Duchenne muscular dystrophy) |
| Adults | Carrier testing (e.g. family history of cystic fibrosis)<br>Presymptomatic testing (e.g. family history of ovarian cancer)<br>Confirmation of diagnosis of adult-onset disorder (e.g. Huntington disease) |

an inherited mutation in the other copy of the gene results in the loss of the normal function of the encoded protein. This gives rise to the tumorigenic effect.

Once the molecular basis of the disease is known in a family, it can be used to identify other family members at risk and to provide prenatal diagnosis for serious conditions or presymptomatic testing (Table 1.3).

DNA testing is now available for many single-gene disorders (Table 1.3). For the current situation regarding a particular condition, see Appendix 1, which gives sources of genetic information, including web-based resources that are continually updated.

Deletions and duplications that are too small to be visualised by light microscopy of chromosomes can be detected using modern specialised PCR-based techniques known as multiple ligation-dependent probe amplification (MLPA) and/or aCGH. These techniques are used for the detection of a wide range of such microdeletions and microduplications, many of which are associated with genetic syndromes (such as those that involve chromosomal regions 1p36 and 9q34). Figure 1.28 shows a high-magnification image of an aCGH slide with thousands of DNA spots, representing different known genomic loci. After hybridisation testing of patient DNA, those spots showing intense red or green indicate either duplications or deletions, depending on the particular prior fluorescent DNA labelling process used. In Figure 1.29,

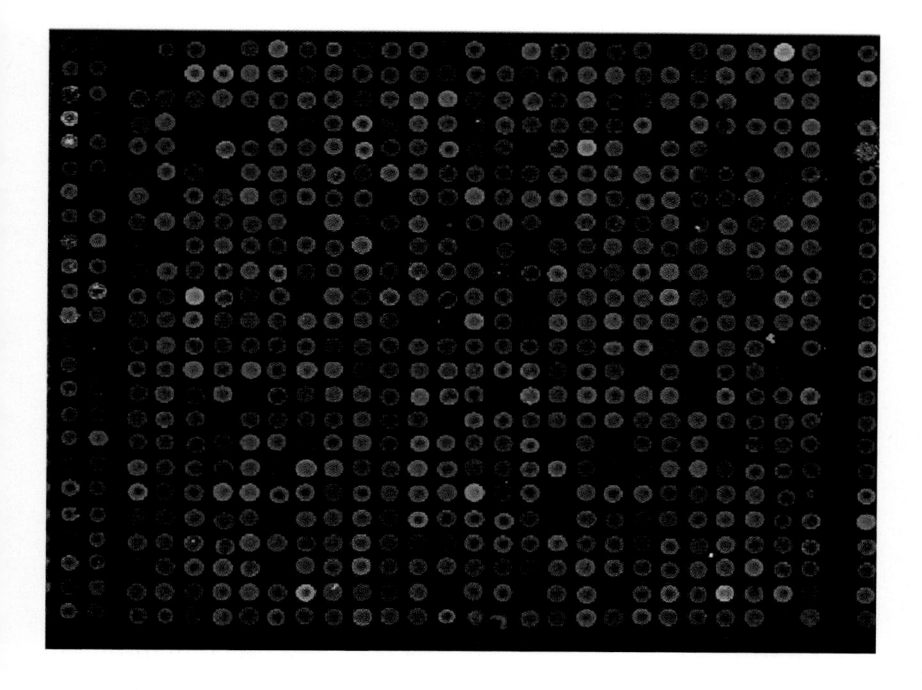

**Figure 1.28** High-magnification image of array comparative genomic hybridisation slide

a microduplication has been detected by this method, as can be seen by the computer prediction to the right of the chromosome diagram (ideogram).

DNA analysis using another PCR-based technique known as QF-PCR is now routinely used to rapidly detect either a trisomy involving chromosome 21, 18 or 13 (causing Down, Edwards and Patau syndromes, respectively) or an abnormal number of sex chromosomes, when one of these aneuploidy syndromes is suspected. Figure 1.30 shows the results of DNA analysis by QF-PCR of DNA from a pregnancy affected by trisomy 21. For each polymorphic marker on chromosome 21 (first three in top left), but not those on chromosomes 18 (top two on right) and 13 (lower panels), there is therefore either a third peak or an abnormal pair of peaks with a 2:1 size ratio. Results from QF-PCR analyses are usually available within 24 hours.

In the future, for diagnostic purposes, newer high-throughput 'massively parallel sequencing' or next-generation sequencing will permit

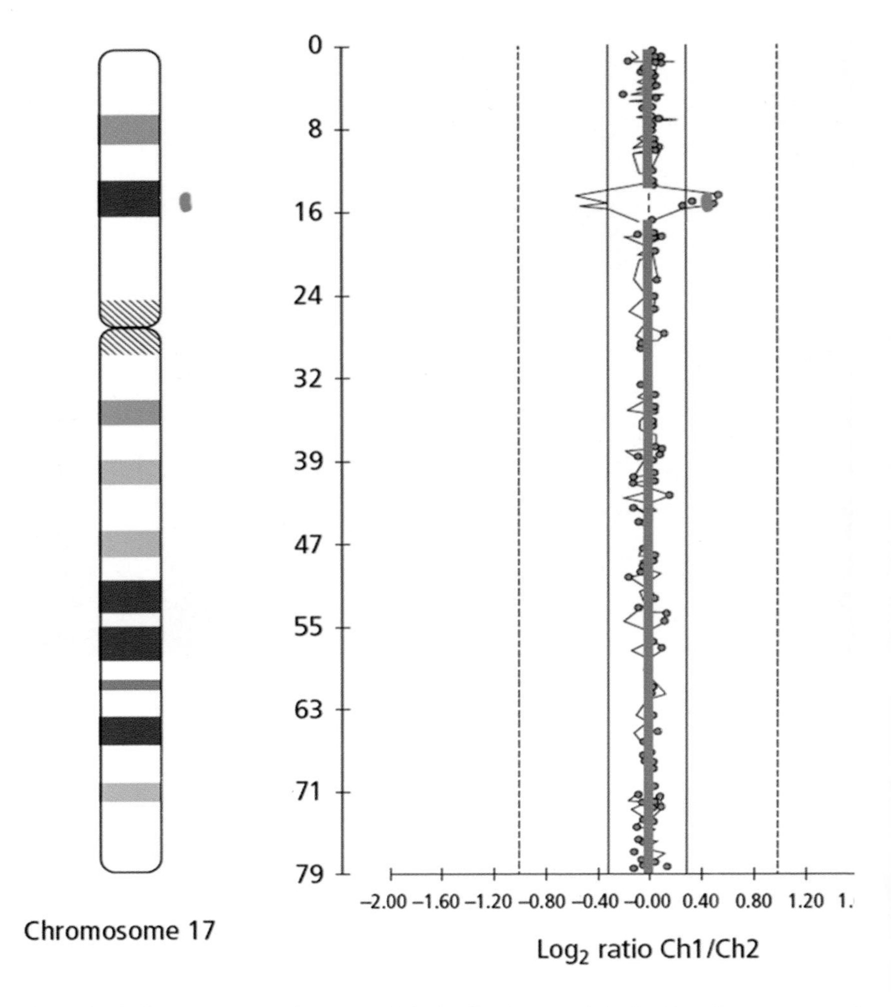


**Figure 1.29** Array comparative genomic hybridisation result indicating the presence of a microduplication on the short arm of chromosome 17

the rapid analysis of multiple genes at once for not only point mutations but also submicroscopic deletions and duplications. Figure 1.31 shows the computer-generated output for next-generation sequencing of part of the *MYH7* gene. Multiple sequence 'reads' are generated (i.e. many horizontal sequence reads from each of the two gene copies), hence the alternative term for this method: massively parallel DNA sequencing. In approximately half of these reads there is a point

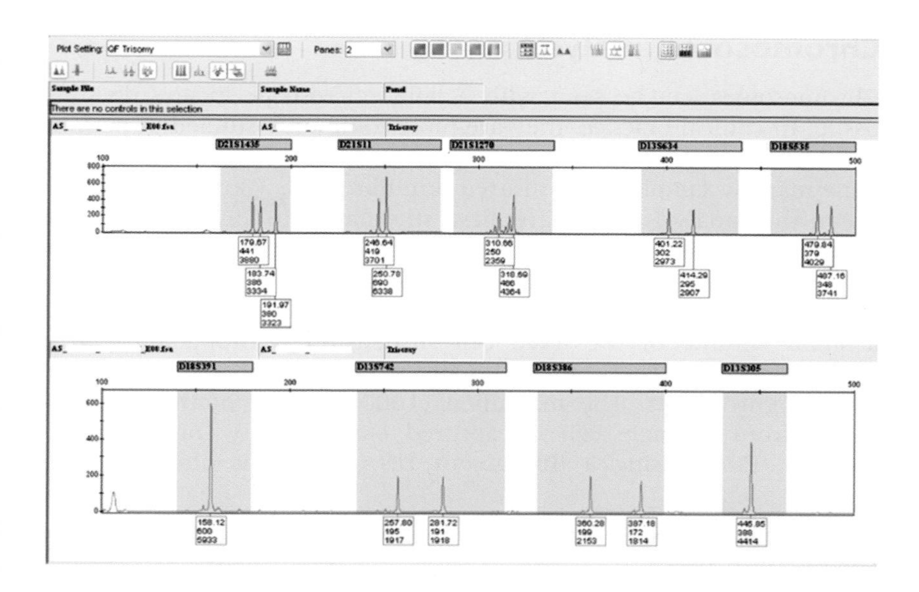


**Figure 1.30** DNA analysis by quantitative fluorescent polymerase chain reaction of DNA from a pregnancy affected by trisomy 21

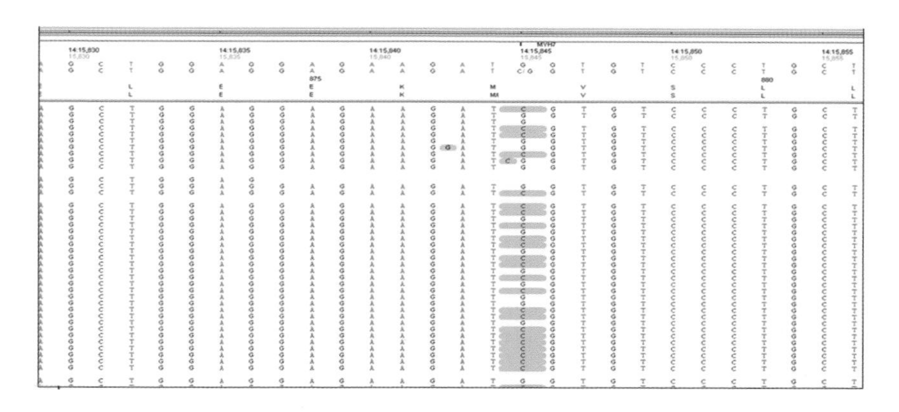

**Figure 1.31** Computer analysis of a section of the *MYH7* gene, following the next-generation sequencing of the DNA of a patient affected by hypertrophic cardiomyopathy. Image kindly provided by Nicola Williams, Head of NHS Molecular Diagnostics, Glasgow, UK.

mutation (with a G nucleotide replaced by a C, as shown highlighted). Thus, this individual, who inherited hypertrophic cardiomyopathy, possesses a heterozygous single nucleotide substitution in the *MYH7* gene.

## Chromosome analysis

Chromosomes can be seen with a light microscope in any dividing tissue. In clinical practice, they are most commonly studied in peripheral blood lymphocytes, amniotic fluid cells, chorionic villus samples, bone marrow samples and cultured skin fibroblasts (Box 1.1).

For this analysis, cells are first stimulated to divide and then arrested in mid-division with colcemid. A hypotonic solution (usually of potassium chloride) is then added, causing the cells to swell but not to burst. Fixative is subsequently added to preserve the cells and a small volume of the fixed cell suspension is dropped on to a microscope slide. This causes the chromosomes to spread out on the slide (Figure 1.32). The magnified (1000x) image of the chromosomes from a single cell is captured electronically for karyotype analysis. FISH, using a fluorescent DNA probe, is often used to

**BOX 1·1 INDICATIONS FOR CHROMOSOME ANALYSIS BY KARYOTYPING**

It should be noted, however, that DNA-based methods are increasingly being used, especially quantitative fluorescent polymerase chain reaction (for aneuploidy) and array comparative genomic hybridisation for the detection of submicroscopic microdeletions and microduplications (in clinically affected cases with dysmorphic features and/or multiple malformations e.g. those marked* below).

Prenatal chromosome analysis:

- Elevated screening risk of Down syndrome
- Parent with a balanced chromosomal translocation
- Previous child with Down syndrome or other autosomal trisomy

Chromosome analysis in neonates:

- Investigation of multiple malformations*
- Confirmation of clinical diagnosis (e.g. Down syndrome)

Chromosome analysis in children:

- Investigation of learning disabilities, dysmorphic features and multiple malformations*

Chromosome analysis in adults:

- Investigation of female or male infertility
- Investigation of recurrent miscarriages
- Family members at risk of carrying a balanced translocation

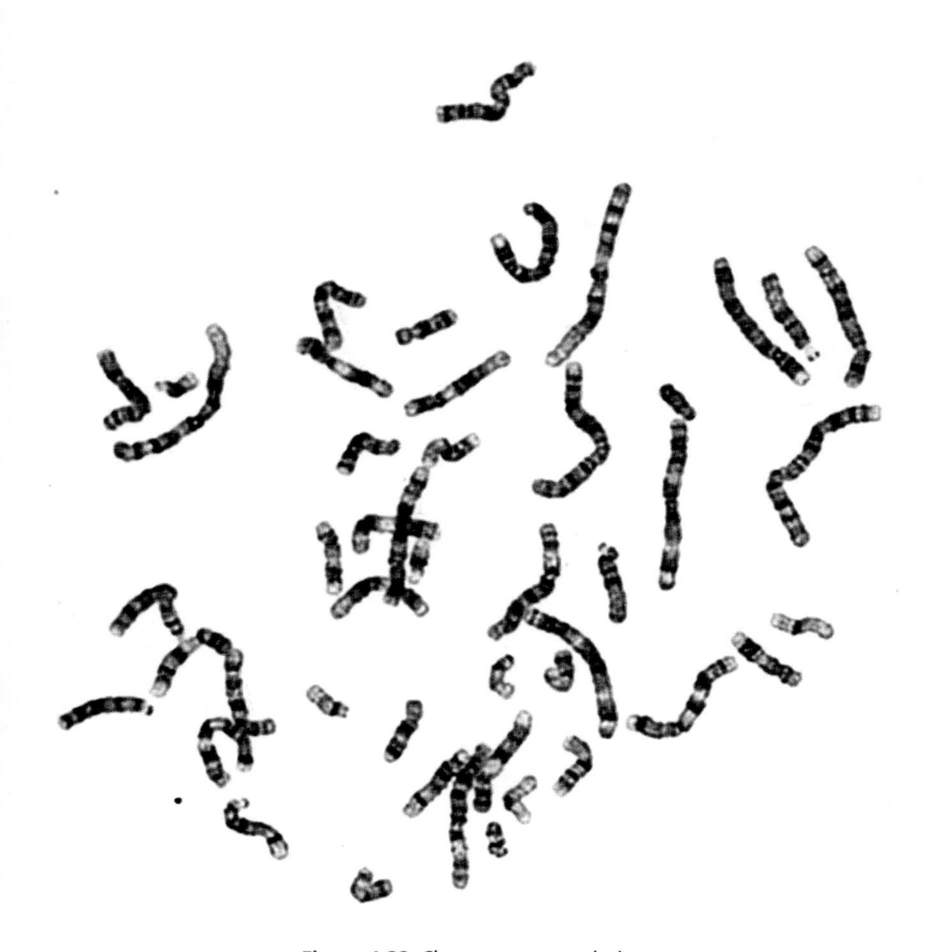

Figure 1.32 Chromosome analysis

confirm the deletion or duplication of a specific chromosome region, which may be suspected from a karyotype, aCGH analysis or clinical signs. DNA-based techniques such as MLPA or aCGH (mentioned above) are being used increasingly to detect small deletions or duplications of less than 4 Mbp in length. In MLPA, the analysis is generally targeted to specific genetic loci.

## Preimplantation genetic diagnosis (PGD)

PGD is a technique that is becoming more commonly undertaken as an alternative to chorionic villus sampling (CVS) or amniocentesis, despite its inconvenience for the individual and its cost. In this technique, after in vitro fertilisation, one or two cells are removed for testing from the embryo when there are just six to ten cells present at day 3. The fetal sex and the presence or absence of a genetic abnormality can then be determined by PCR or by FISH. Unaffected embryos can then be implanted into the uterus to allow the pregnancy to continue. A great many embryos have now been successfully tested by PGD for well over 200 single-gene disorders, including fragile X syndrome, cystic fibrosis and myotonic dystrophy, and to test for unbalanced chromosome abnormalities that can result from a chromosomal translocation in a parent. The technique is improving, but in a high proportion of PGD attempts it does not result in a successful birth. Furthermore, the procedure is generally found to be physically and emotionally demanding. As a consequence of these various factors, only a minority of couples select PGD rather than alternative reproductive options.

## Cell-free fetal DNA testing

An increasing number of centres in the UK are offering fetal sex determination by the analysis of cell-free fetal DNA (cffDNA) in maternal plasma. This relies on the presence in the plasma of circulating DNA molecules, some of which are fetal in origin. Using sensitive molecular genetic techniques it is possible to analyse this DNA and to detect the presence of gene sequences that are either derived from the father or that have newly arisen. Thus, it is possible to determine whether the fetus is male or female, without the need for either CVS or amniocentesis, by testing for male-specific sequences. This is likely to be a particularly helpful method for the pregnancies of women who are carriers of X-linked recessive conditions in order to identify male pregnancies. One method for such an analysis is the realtime PCR assay (i.e. with continuous fluorescence measurement during the PCR reaction) of the male *SRY* gene sequence. These analyses can be challenging, however, on account of the fact that in early pregnancy, the fetal DNA constitutes less than 10% of the free circulating DNA in the maternal serum.

This non-invasive prenatal diagnosis (NIPD) method is also being developed for the detection of RHD in RhD-negative mothers and the diagnosis of some monogenic disorders including achondroplasia. In the future, despite the technical challenges involved, it is likely that it will be possible to carry out NIPD by analysing cffDNA in the maternal circulation for the diagnosis of trisomy 21 and other aneuploidies, particularly in high-risk pregnancies. This might be undertaken by next-generation sequencing (massively parallel sequencing). The use of such a method has in fact already been evaluated for the diagnosis of trisomy 21, in high-risk pregnant women, with highly promising results (see Chiu et al. 2011 in Further reading).

## Referral for genetic assessment and counselling

When a patient asks for genetic advice or if a healthcare professional suggests that a disorder might have a genetic basis and seeks advice from a medical geneticist about diagnosis and genetic advice for the family, the usual method of communication is by referral letter. The affected person who brings the family to medical attention is called the 'proband' and the person who is seeking advice, and who may or may not be affected, is called the 'consultand'. Contact details for the UK network of regional genetics centres are provided online (see p. 75 for url).

When a family is seen at the genetic clinic a detailed family tree is drawn and the diagnosis is confirmed in each affected person. Depending on the condition, genetic tests may need to be performed or the patient may need to be referred to another specialist. Apparently healthy members of the family may need to be examined in case they show mild signs of the disorder.

Genetic disorders may affect any organ system and a condition might have multiple component features. Recognition that these component features are interlinked (that is, they are a syndrome) is crucial for clinical management. Problems can arise when individual specialists concentrate on single components and no one sees the whole picture. The process is not helped by the variability of many syndromes. The classic textbook descriptions are uncommon and most patients do not have a 'full house' of clinical features.

Genetic counselling will cover all aspects of the condition including the prognosis for affected persons, carrier risks for healthy persons, recurrence risks and reproductive options. Accurate diagnosis is of paramount importance for meaningful genetic counselling, so counselling should never precede the steps involved in diagnosis as outlined above. Ideally, both parents should be counselled and adequate time allowed in an appropriate setting. Few couples can be counselled in less than 30 minutes and neither the corner of a hospital ward nor a crowded clinic room is adequate. It is inappropriate to counsel too soon after bereavement or after the initial shock of a serious diagnosis. The depth of explanation needs to be matched to the educational background of the couple and compared with other information they have gleaned from medical sources, parent support groups and internet sites. Counselling must be nonjudgemental and nondirective. The aim is to deliver a balanced version of the facts that will permit those who are seeking advice to reach their own decisions in terms of their reproductive future.

The pedigree number is the same for all members of a family and each person's records are stored in the (written or electronic) family case notes that are indexed under this number. This facilitates interpretation of the family tree and the results of any genetic tests. It also makes it easier to ensure that all at-risk family members have been offered genetic advice. Although many families choose to come to the clinic as a group, the principles of medical confidentiality should still be observed. Many people would not wish relatives to know the results of their personal genetic tests. Following the consultation a letter summarising the information is sent to the consultands with copies to the referring clinician and the consultand's general practitioner. A medical version of the same information is also sent to the referring clinician and the consultand's general practitioner.

Consultands often feel guilty or stigmatised and it is important to recognise and allay this. Common misconceptions about heredity may also need to be dispelled (Box 1.2).

For certain conditions, such as balanced translocations, autosomal dominant traits and X-linked recessive traits, an extended family study

## BOX 1·2 COMMON MISCONCEPTIONS ABOUT HEREDITY

- Presence of only one affected person in the family means that the condition is not inherited (or genetic).
- Presence of other affected family members means that a disorder is always genetic.
- If only males or females are affected in the family then only this sex can be affected.
- Any condition present at birth (i.e. congenital) must be genetic.
- Upsets, mental and physical, of the mother in pregnancy cause malformations.
- A one in four risk means that the next three children will be unaffected.
- Confusion of odds, fractions and percentage risks.
- All genetic disorders and their carrier state can be detected by chromosomal analysis.
- All genetic disorders and their carrier state can be detected by DNA analysis.

will be required and it is useful to enlist the aid of the consultands in approaching other family members at risk.

Many consultands can be fully counselled at one sitting, but some will require follow-up sessions. If new opportunities arise (such as an improved carrier or prenatal diagnostic test), consultands can be contacted and offered a return appointment.

# Section Two

## Common genetic problems in obstetric and gynaecological practice

# Common genetic problems in obstetric and gynaecological practice

## Introduction

This section highlights the common situations in obstetric and gynaecological practice where there are genetic implications.

## Genetic causes of infertility

In the UK, about one in six or seven couples is affected by involuntary infertility at some point (although many of these will achieve pregnancy naturally, in time). Investigation of both male and female primary infertility should include chromosome analysis. An additional investigation in the male is DNA testing for milder combinations of cystic fibrosis mutations. These result in absence of the vas deferens and infertility but no other features of cystic fibrosis.

One of the most common chromosomal causes of male infertility is Klinefelter syndrome, which is usually not diagnosed until adulthood. In addition to infertility, patients may show mild undermasculinisation and gynaecomastia. Intelligence and general lifespan are usually within the normal range. There is an increased risk (7%) of diabetes mellitus and also of male breast cancer. Management will include assessment for hormonal replacement and screening for complications.

The diagnosis of Klinefelter syndrome is established by chromosome analysis, which reveals 47 chromosomes in total, with an additional X chromosome (Figure 2.1). Despite hormone replacement, without assisted conception, men with Klinefelter syndrome are azoospermic and infertile and so there is not generally a genetic implication for their offspring. With assisted reproduction techniques, such as microsurgical testicular sperm extraction (micro-TESE) coupled with intracytoplasmic sperm injection (ICSI), however, affected men can father children. The genetic risks to the offspring of these men are currently under investigation and may be much lower than expected (see Fullerton et al. 2010 in Further reading).

One of the most common chromosomal causes of primary female infertility is Turner syndrome. The diagnosis of Turner syndrome is

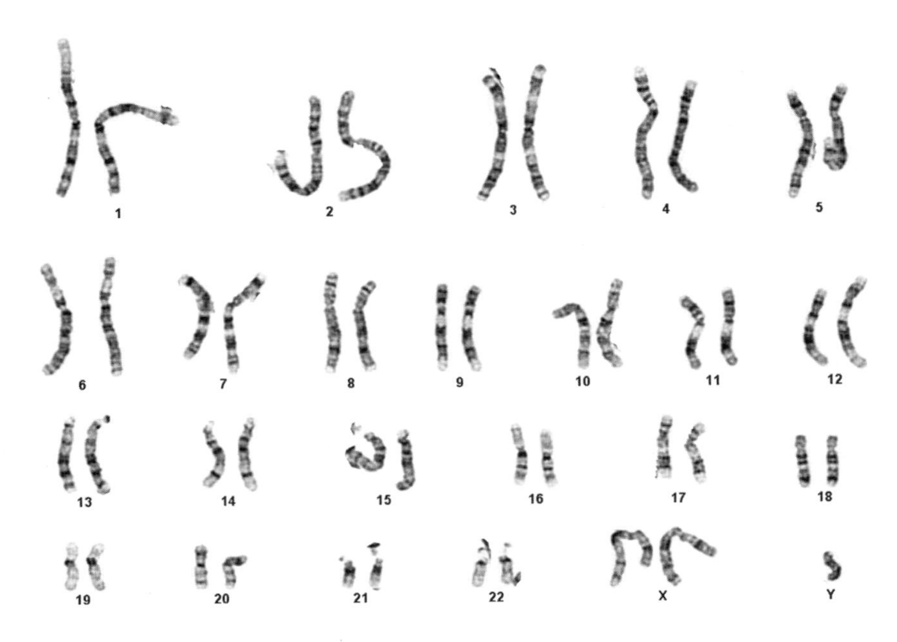


**Figure 2.1** Karyotype of Klinefelter syndrome: 47,XXY

established by chromosomal analysis. This usually reveals only 45 chromosomes in total, with a single X chromosome (see Figure 1.6, page 8). There are several other less common chromosomal patterns that also result in Turner syndrome.

The frequency of Turner syndrome at conception is estimated to be as high as 1%. The majority of these pregnancies spontaneously miscarry and most are undiagnosed. Clinical suspicion during pregnancy may be raised by the ultrasound appearance with nuchal oedema in a female fetus (Figure 2.2). Surviving fetuses may occasionally be diagnosed in the newborn period if they show the combination of puffy hands and feet (Figure 2.3), deepset nails and neck webbing, but most remain undiagnosed until they are investigated for childhood short stature or adult infertility. The classic depiction of an adult with Turner syndrome with marked neck webbing, a broad chest and a wide elbow-carrying angle relates to only a minority of women and many are normal apart from short stature and infertility. Intelligence is normal. Short stature is universal, with an untreated adult height of about 1.5 m. Infertility is usual but not invariable and 10% have a spontaneous menarche. Women with Turner syndrome have an increased risk of congenital heart disease, especially coarctation of the aorta and of systemic hypertension.

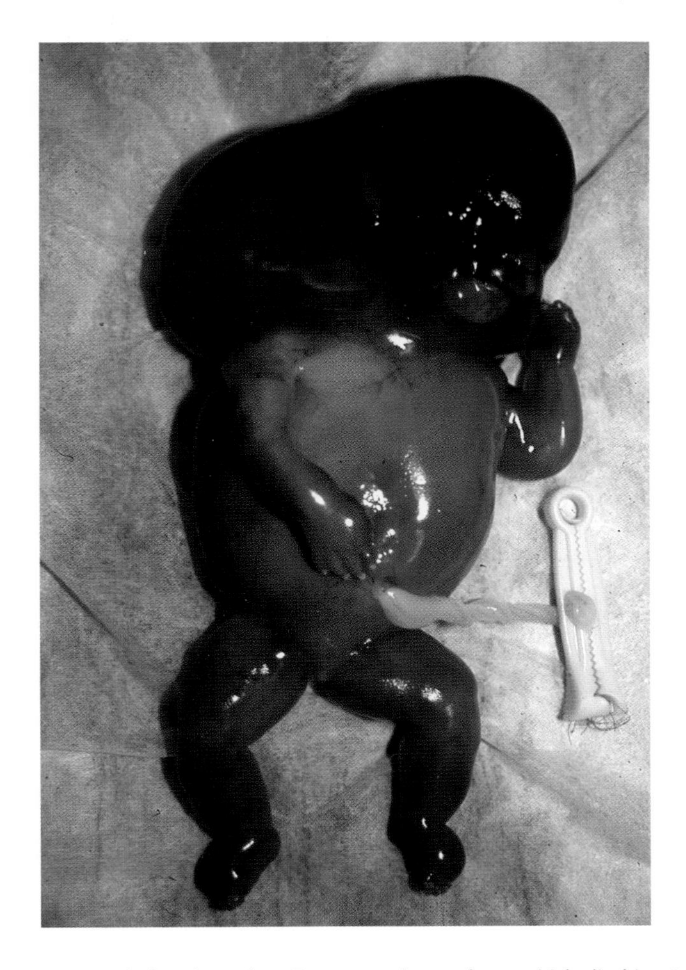

**Figure 2.2** Nuchal oedema in a Turner syndrome fetus which died in utero

Management of women with Turner syndrome includes childhood use of growth-promoting agents and replacement of female sex hormones.

## Genetic causes of recurrent miscarriages

At least one in six recognised pregnancies ends as a miscarriage and so, by chance alone, several miscarriages are not uncommon. Genetic investigation is appropriate for three or more unexplained first-trimester miscarriages as, in this situation, one parent will be found to be a carrier of a chromosomal translocation in 5% of investigated couples.

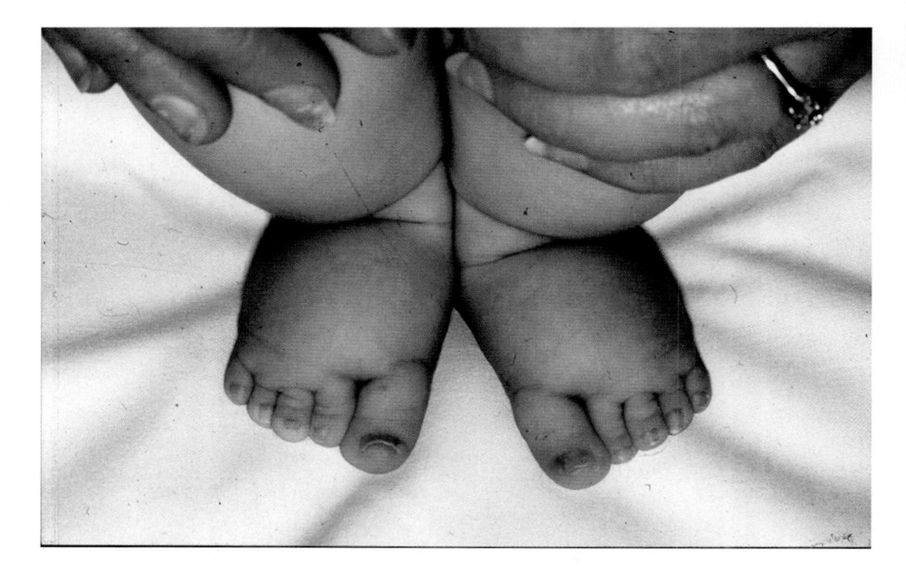

**Figure 2.3** Puffy feet in a neonate with Turner syndrome

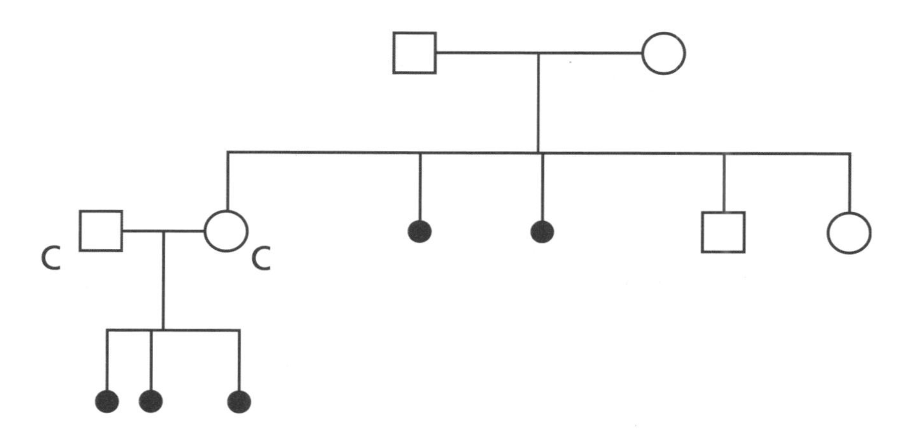


**Figure 2.4** Pedigree showing recurrent miscarriages

In the family shown in Figure 2.4, the couple was investigated in view of their recurrent miscarriages. Parental chromosomes revealed a normal karyotype in the father but the mother possessed a translocation between chromosomes 8 and 9 (Figure 2.5). Genetic counselling of these parents would cover how the translocation arose and its implications for the carrier's health, for the next pregnancy and for other

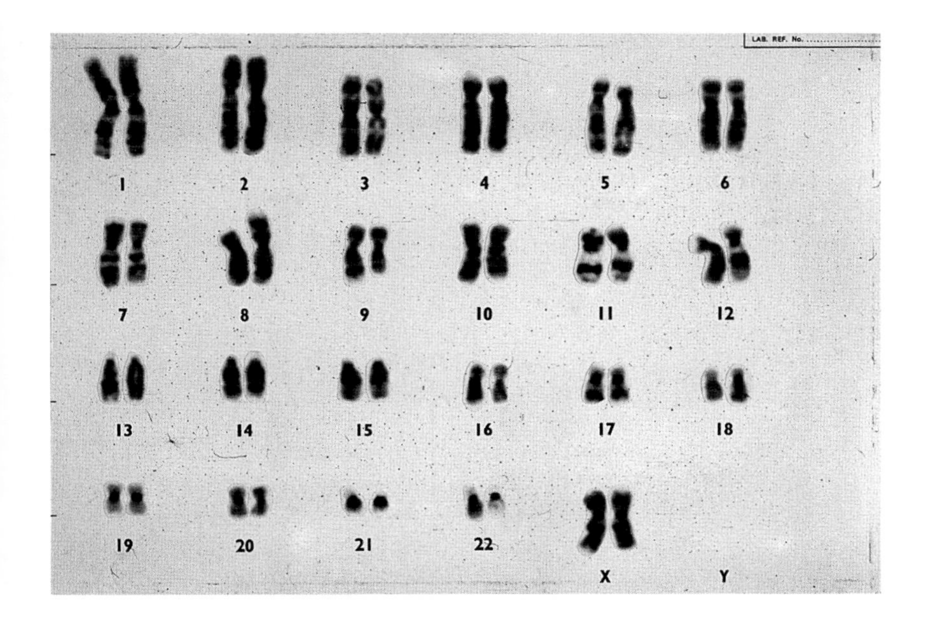


**Figure 2.5** Karyotype with a balanced translocation between chromosomes 8 and 9

family members. When chromosomes break, the ends are sticky and normally reunite with no clinical effect. If, however, more than two chromosomes break at the same time, there is a chance that the sticky ends will unite in a new arrangement. For this family, breaks have occurred in chromosomes 8 and 9, either at the time the mother was conceived or in an earlier generation. Generally, genes are not damaged at the chromosomal breakpoints and they function normally despite the new arrangement. This is termed a balanced translocation, as the carrier has no loss or gain of genetic material. The carrier mother can thus be reassured that the fact that she is a carrier of a translocation should not influence her general health or lifespan.

The genetic implications for this mother relate to reproduction. Her partner passes on one copy of chromosome 8 and one of chromosome 9. She must pass on either the normal copy or the translocated copy of each of these chromosomes (Figure 2.6). She can thus have children with normal chromosomes, children with the same balanced translocation as herself and children with various gains or losses of chromosomal material. These gains or losses may result in an early spontaneous miscarriage, but this is not invariable and a liveborn child with such an imbalance would be likely to have multiple congenital malformations and major learning difficulties. Where a translocation is identified during the investigation

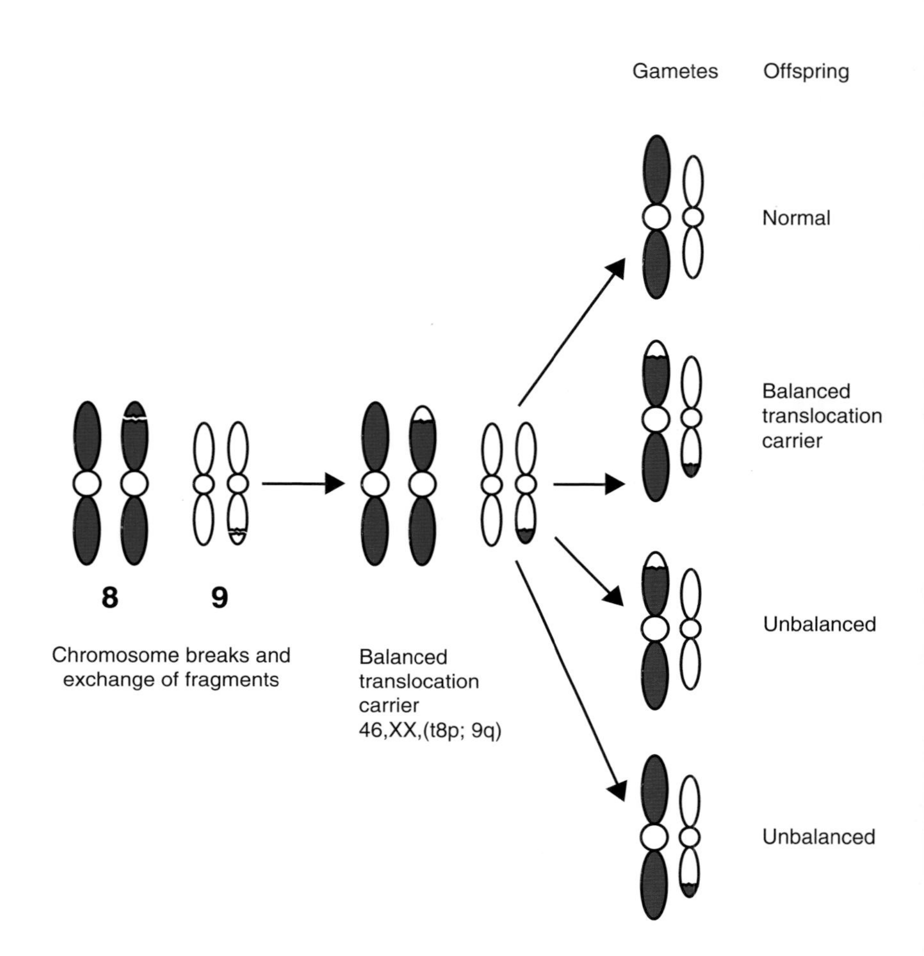


**Figure 2.6** Types of gamete for a carrier of an 8;9 chromosomal translocation

of recurrent miscarriage, the additional risk for a couple of having a live-born disabled child depends on the precise chromosomal breakpoints and the extent to which the rearrangement can interfere with meiosis. This requires careful study by the geneticist with the help of the cytogenetics laboratory. Prenatal diagnosis and chromosome analysis should be discussed. The couple will need to consider the relative advantages and disadvantages of chorionic villus sampling (CVS).

Amniocentesis is generally performed from 16 weeks of gestation onwards and a karyotype result is usually available in 2 weeks. As a consequence, couples faced with a serious chromosomal imbalance in the fetus need to consider a late termination of pregnancy. CVS can be offered from 10 weeks of pregnancy and again a karyotype is generally available within 2 weeks. Hence, if requested, a first-trimester termination of pregnancy is then possible. The disadvantages of CVS are its more restricted availability and the fact that the test itself carries a higher risk of miscarriage. This risk is commonly quoted as 2%, which means that, for every 100 tests, two pregnancies are lost. For amniocentesis, the commonly quoted risk of miscarriage attributable to the procedure is 1%.

As for other genetic disorders, the medical responsibility of the clinician who makes the diagnosis relates to the whole family and not just one patient. In the family shown in Figure 2.4 (page 46), the first step is to offer to test the mother's parents' chromosomes. If these are normal then the translocation has arisen at the time the mother was conceived and risks to siblings and cousins will be very low. In this family, the mother's father was found to carry the same balanced translocation and testing of her brother and sister and the father's relatives should be offered. In practice, investigation of the extended family will usually be undertaken by the local regional genetics service. See Appendix 1 for the web address of the directory of UK genetics centres.

## Elevated maternal screening risk

### DOWN SYNDROME

In the absence of a family history, the risk of Down syndrome in a pregnancy can be calculated from a combination of the mother's age and the results of biophysical tests.

The risk or probability of an event ranges from one (100% or always happens) to zero (0% or never happens). On the basis of maternal age alone, the risk of Down syndrome at birth varies from one in 1500 for a 20-year-old mother to one in 28 for a 45-year-old mother (Table 2.1).

In the first trimester, analysis of maternal biochemical markers - usually a combination of maternal serum free β human chorionic gonadotrophin (FβhCG) and pregnancy-associated plasma protein A (PAPP-A) - can provide a relative risk figure for the presence of Down syndrome in that pregnancy. This is combined with the maternal age-related risk (from Table 2.1) and with a further independent risk based on the sagittal thickness of the nuchal fold (nuchal translucency) on ultrasound examination. In general, in the first trimester, FβhCG and nuchal translucency are increased in pregnancies affected by Down syndrome, whereas PAPP-A is reduced.

For example, a 30-year-old mother has an age-related risk at birth of one in 900. Her level of FβhCG is found to be 1.8 MOM, giving a likelihood ratio of 1.5 (Figure 2.7). Her level of PAPP-A is 0.5 MOM and this gives a likelihood ratio of 2.0 (Figure 2.8). Her fetal nuchal

| Table 2.1 Frequency of Down syndrome at birth in relation to maternal age | |
|---|---|
| *Maternal age (years)* | *Approximate frequency of Down syndrome at term* |
| 20 | 1:1528 |
| 25 | 1:1351 |
| 30 | 1:909 |
| 35 | 1:384 |
| 37 | 1:240 |
| 39 | 1:150 |
| 40 | 1:112 |
| 41 | 1:85 |
| 43 | 1:50 |
| 45 | 1:28 |

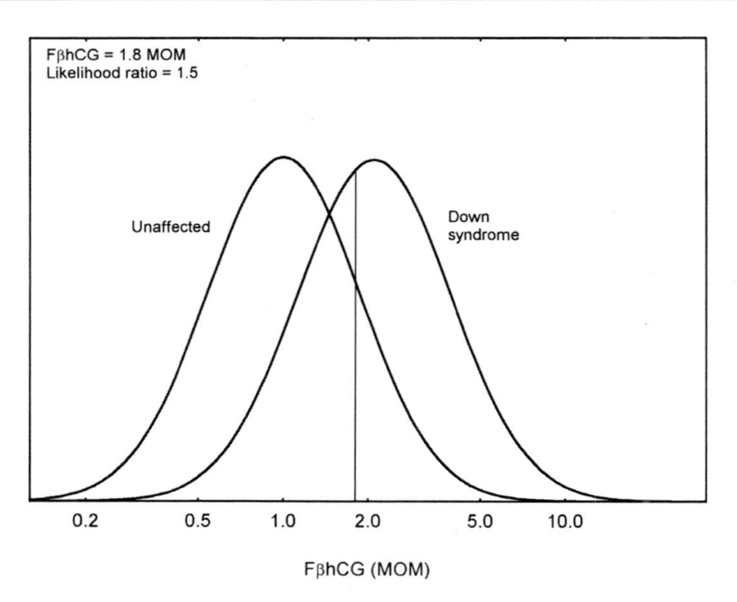


**Figure 2.7** Use of maternal serum FßhCG expressed in multiples of the median to give a likelihood ratio for Down syndrome in the first trimester

thickness measurement is 1.8 MOM and this gives a likelihood ratio of 4.0 (Figure 2.9). The biochemical markers are independent of each other and of the nuchal thickness (and the maternal age) and so are combined by multiplication. Her combined risk is thus approximately one in 75 (1:900 x 1.5:1 x 2:1 x 4:1). This first-trimester combined

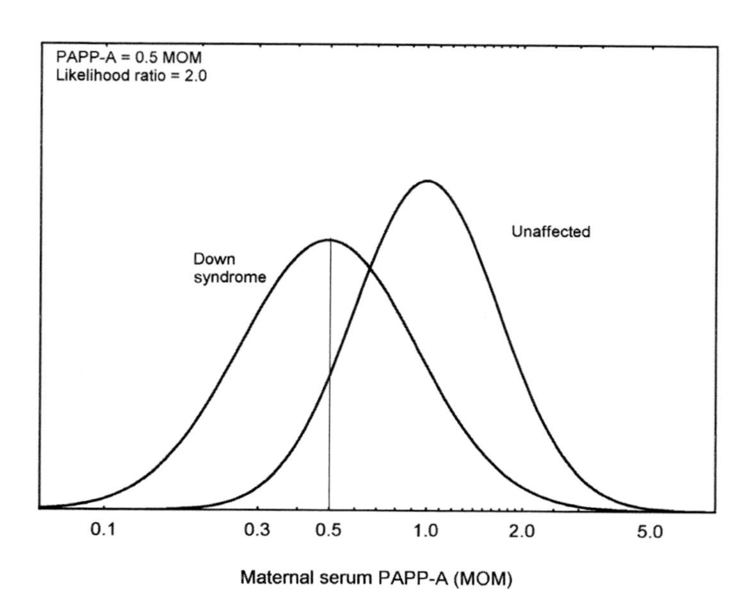


**Figure 2.8** Use of maternal serum PAPP-A expressed in multiples of the median to give a likelihood ratio for Down syndrome in the first trimester

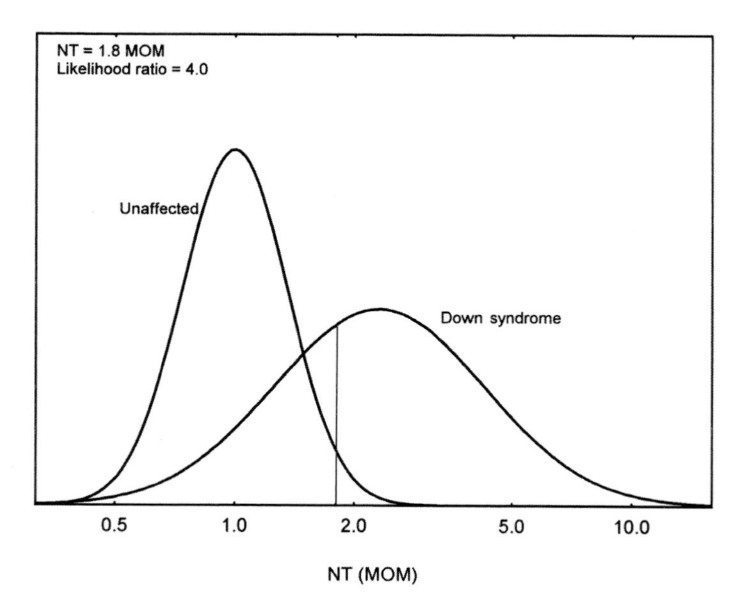


**Figure 2.9** Use of sagittal thickness of the nuchal fold expressed in multiples of the median to give a likelihood ratio for Down syndrome in the first trimester

(ultrasound and biochemical screening) test is now the recommended screening method for Down syndrome in the UK, undertaken at $11^{+2}$ to $14^{+1}$ weeks of gestation (although the blood sample itself can be taken between $10^{+0}$ and $14^{+1}$ weeks).

For women unable to have the first-trimester combined test because of, for example, late booking, biochemical screening may be offered in the second trimester. For analysis of maternal biochemical markers at that stage, a variable combination of FβhCG or, more usually, total hCG, alphafetoprotein (α-FP), unconjugated estriol 3 (UE3) and inhibin A can also provide a relative risk figure for the presence of Down syndrome in that pregnancy. For instance, quadruple screening may be used, involving α-FP, hCG, UE3 and inhibin A. Again, the risks are independent of each other and can thus be combined by multiplication. In the second trimester, maternal serum α-FP and UE3 levels are generally reduced in pregnancies affected by Down syndrome, while the levels of inhibin A, hCG and FβhCG are usually increased (see Table 2.2).

The level of risk at which CVS or amniocentesis is offered varies but is commonly one in 150 (in line with the recommendations of the UK National Screening Committee). Thus, a woman with a combined risk of one in 100 would fall in the high-risk category and be eligible for prenatal chromosome analysis.

In communication of these risks to the patient, several difficulties need to be anticipated and covered. First, this is a screening test and not a diagnostic test. A diagnostic test confirms or refutes a diagnosis in an individual and should be approximately 100% reliable. By contrast, a screening test aims to identify high- and low-risk groups and has to be the first step when universal use of the diagnostic test is impractical. The high-risk group

**Table 2.2 Maternal serum proteins and their variation in pregnancies affected by trisomies**

| *Protein level in maternal serum* | *Trisomy 21* | *Trisomy 18* | *Trisomy 13* |
|---|---|---|---|
| **First trimester** | | | |
| FβhCG | High | Low | Low |
| PAPP-A | Low | Low | Low |
| **Second trimester** | | | |
| α-FP | Low | Low | Normal/high |
| hCG | High | Low | Normal |
| UE3 | Low | Low | Low* |
| Inhibin A | High | Normal | High* |

Data from Aitken et al. 2007. * Based on small numbers. α-FP = alphafetoprotein; FβhCG = free beta human chorionic gonadotrophin; PAPP-A = pregnancy-associated plasma protein A; UE3 = unconjugated estriol 3.

will contain most but not all affected pregnancies but will also include many normal pregnancies. The low-risk group will still contain some affected pregnancies but most will be normal pregnancies. As only those in the high-risk group are offered the diagnostic test (in this case fetal chromosome analysis), not all affected pregnancies can be detected and the sensitivity will be less than 100%. In second-trimester screening programmes the sensitivity or detection rate is about 70-75%, which means that the false negative rate is about 25-30%. In first-trimester screening the sensitivity or detection rate is around 85-90%.

The overall chance that a woman who has a screening test will be offered a diagnostic test is around one in 30. If a woman then has a diagnostic test, the overall chance of finding Down syndrome is at least one in 28.

The second difficulty relates to understanding risks. Some people fully understand the meaning of a one in 100 risk but many need further explanation with, for example, alternative expressions as percentages or as odds of normality (e.g. 99 times out of 100 will be normal). It may also be helpful to translate risks into revised maternal age risks. Thus a 40-year-old mother with a combined risk of one in 380 equates to that for a 35-year-old mother and she may decide against fetal chromosome analysis. Some people may find it helpful to know how close they are to the threshold between high and low risk.

The third difficulty relates to the parents' perception of Down syndrome. Maternal screening is only appropriate in the absence of a family history of Down syndrome and in that situation the parents will have little or no direct knowledge about the condition. The main problem for children with Down syndrome relates to major learning difficulties. Typically, children will learn to walk and talk but will remain dependent on their parents or carers. They have minor physical characteristics that aid clinical diagnosis. About one-third will have major congenital heart malformations that may need surgical correction. In the absence of congenital heart disease or with successful correction, the average lifespan is still reduced because of other complications including dementia and leukaemia (1%).

It is essential that parents offered antenatal screening programmes have adequate information about the condition and the limitations of the test in order to make an informed choice. Also, women having fetal karyotyping as a prenatal diagnostic test should be informed of the possibility of an unexpected chromosomal abnormality being detected, e.g. Klinefelter syndrome (see Case 9 in Section 3).

## NEURAL TUBE DEFECT

In the absence of a family history of a neural tube defect (NTD; i.e. anencephaly, spina bifida or encephalocele), the risk for a pregnancy can be assessed by detailed fetal ultrasound scanning or by maternal serum screening.

Ultrasound scanning in the first trimester (e.g. to confirm the stage of gestation) can detect almost all cases of anencephaly but it is difficult to visualise spina bifida at that stage.

Detailed ultrasound scanning in the second trimester for fetal malformations (e.g. of the central nervous system, kidneys, heart and limbs) is now widely available in many countries. Although the published data show large variations, the sensitivity of NTD detection has been found to be higher than for maternal serum α-FP (MSα-FP) screening (e.g. see Dashe et al. 2006 in Further reading). This has resulted in ultrasound becoming the recommended method of screening for NTD in the UK.

Prior to the widespread availability of second-trimester detailed ultrasound scanning, maternal serum screening was undertaken instead. Open NTDs lack a covering of skin over the lesion and fetal α-FP can thus leak into the amniotic fluid and then reach the maternal circulation. In normal pregnancies, MSα-FP levels start to rise from 10 weeks and peak at around 32 weeks, before falling to term. First-trimester screening for NTDs using MSα-FP is not possible, as the levels are not elevated until mid-pregnancy. Thus, measurement of MSα-FP was undertaken at 15–20 (optimum 16–17) completed weeks of gestation (Figure 2.10). If the level was above the 95th centile (equivalent to 2 MOM for that gestation), detailed fetal ultrasound was indicated, as there would be at least a 2% chance of a NTD.

For MSα-FP measurement, as emphasised for Down syndrome screening, several difficulties need to be anticipated and covered in communication with the parent. First, MSα-FP measurement is a screening method that aims to identify a high-risk group who can then be offered the diagnostic test for a NTD (detailed ultrasound scanning). The sensitivity or detection rate depends on the type of defect. Anencephaly (Figure 2.11) is generally 'open', with no skin covering, and the area of exposed tissue typically gives extremely high levels of MSα-FP. The sensitivity is thus virtually 100% for anencephaly. Spina bifida, by contrast, is 'open' in just 80–85% of cases (Figure 2.12). The 'closed' skin-covered lesions (Figure 2.13) cannot be detected using this method of screening. Moreover, the size of open spina bifida varies from extensive lesions that can cause elevations of MSα-FP similar to those seen for anencephalic pregnancies to small lesions, which may not produce abnormal elevations of MSα-FP. The sensitivity of MSα-FP screening for open spina bifida was around 88%. Thus, 12% of open spina bifida would be assigned to the low-risk group and missed.

The second difficulty relates to understanding risk figures. The overall risk for the high-risk group of having a NTD was one in 50 but, in

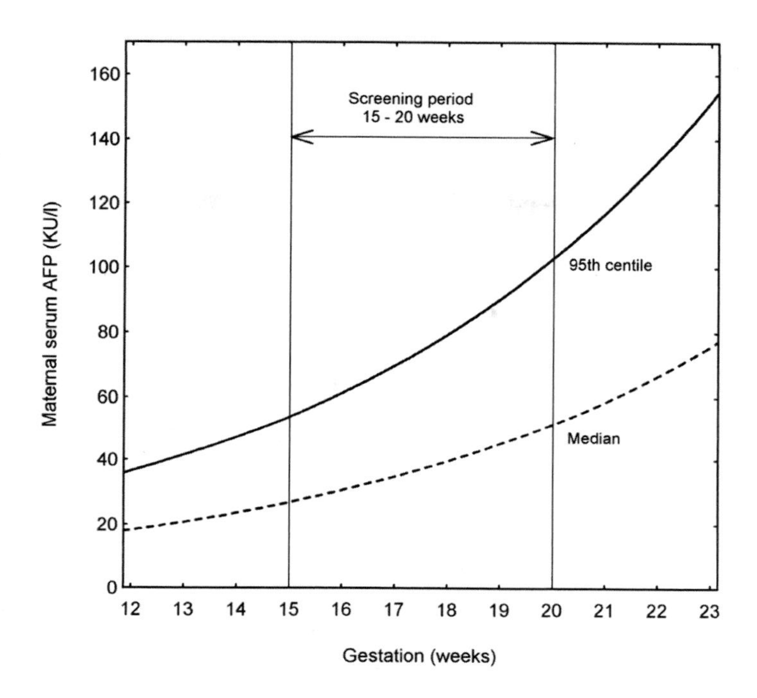


**Figure 2.10** MSα-FP centiles for the screening period

practice, numerical risks were usually avoided and the mother was assigned as high or low risk.

An additional difficulty (following the detection of NTD by either method) relates to providing the couple with an outlook for the affected pregnancy. Stillbirth or neonatal death is invariable for anencephaly but the prognosis for spina bifida is more difficult to predict. For open spina bifida, even with surgery within 24 hours, a high proportion of those that survive are severely disabled. Babies with gross paralysis of the legs, thoracolumbar or thoracolumbar–sacral lesions, kyphoscoliosis, hydrocephalus at birth or associated abnormalities have a particularly poor prognosis.

It is essential that women offered antenatal screening programmes have adequate information about the condition and the limitations of the test in order to make an informed choice.

## Cystic fibrosis

Cystic fibrosis is a relatively common cause of childhood morbidity and mortality. The clinical severity is variable, with problems mainly relating to pancreatic insufficiency and chronic lung disease secondary to

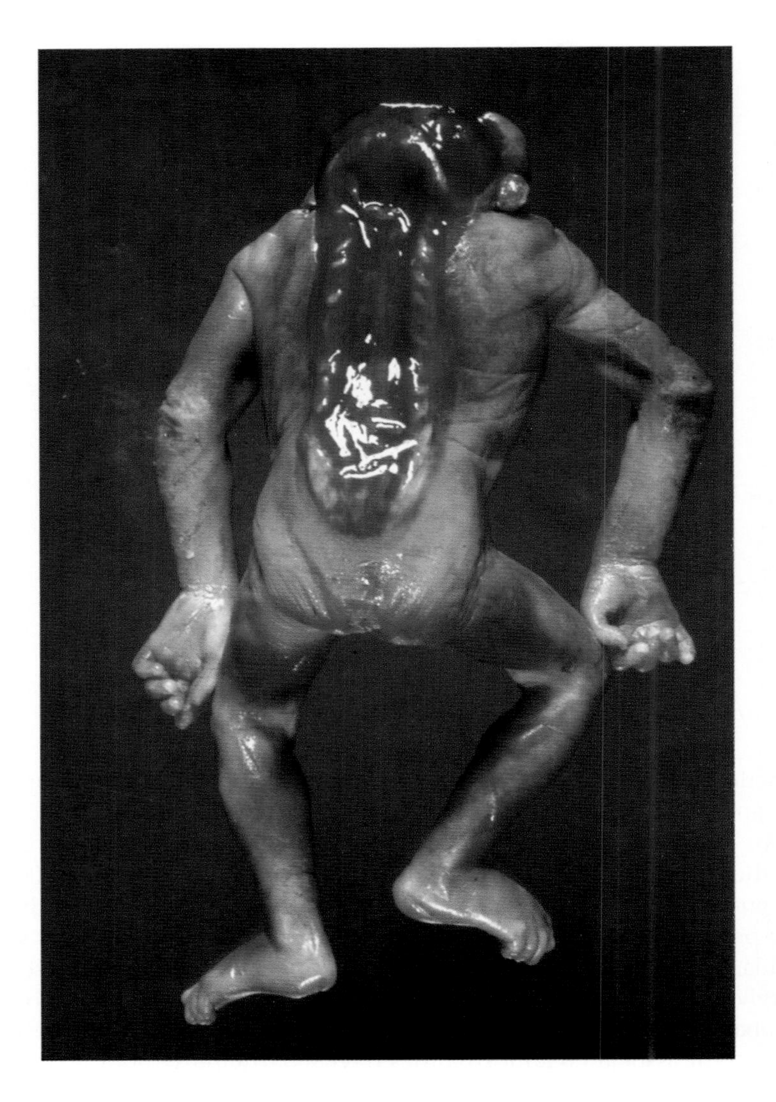

**Figure 2.11** Anencephaly and extensive open spina bifida

recurrent infection. The prognosis has improved with advances in treatment such that the overall median survival is now between 35 and 40 years (Cystic Fibrosis Foundation, www.cff.org/AboutCF/).

Approximately one in 2000 pregnancies is affected. The affected pregnancies have inherited an underactive *CFTR* gene from each parent. The parents are carriers (heterozygous) with one normal *CFTR*

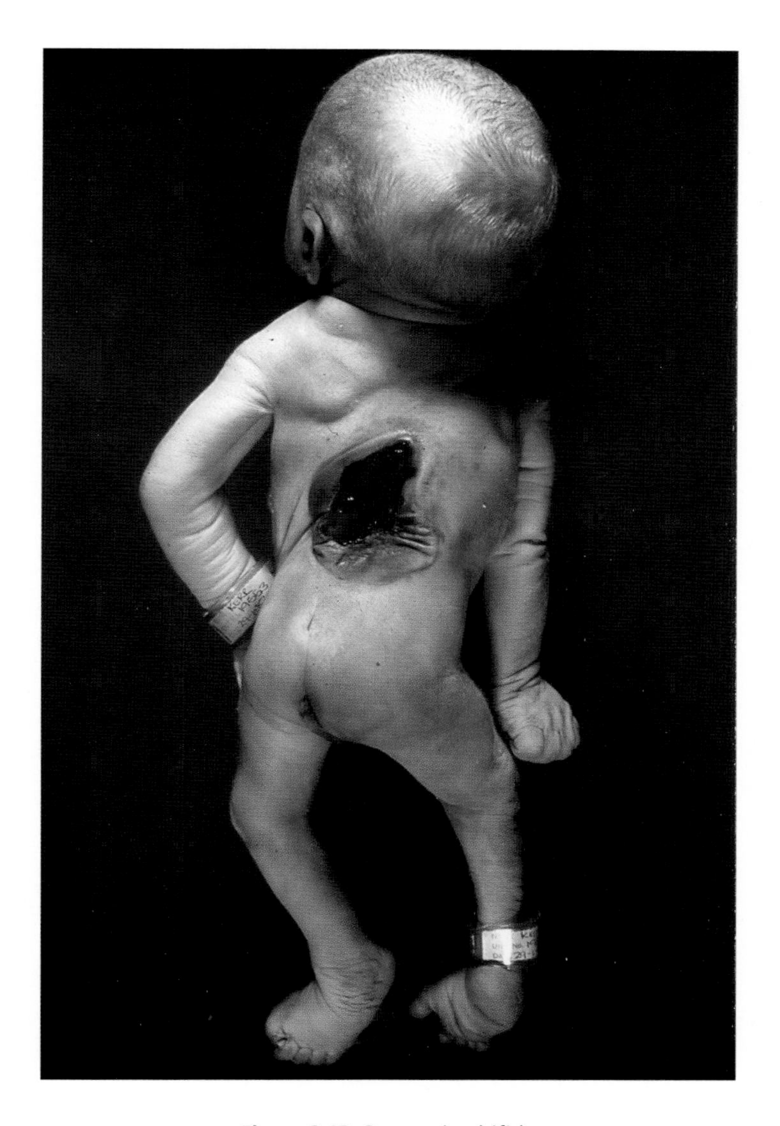

**Figure 2.12** Open spina bifida

gene on one copy of chromosome 7 and one underactive *CFTR* gene on the opposite copy of chromosome 7. For these carrier parents, on average one in four of their pregnancies will inherit two copies of the underactive cystic fibrosis gene and therefore result in an affected child (see Figure 1.12, autosomal recessive inheritance, page 15).

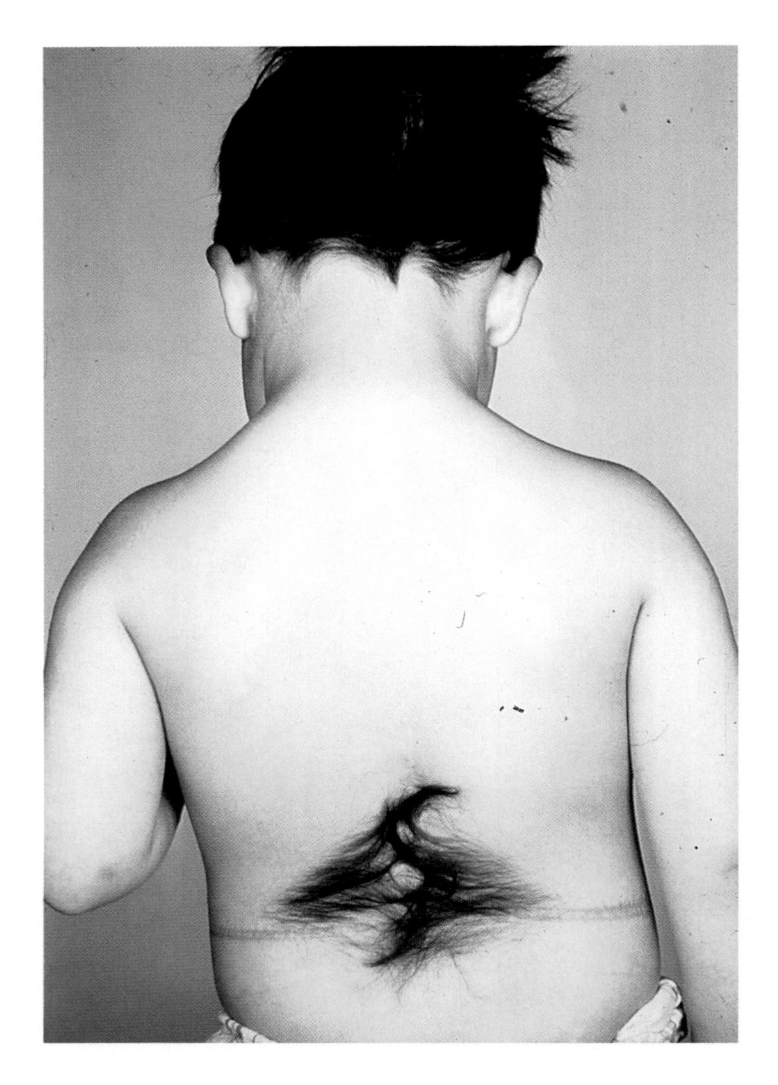

**Figure 2.13** Closed spina bifida

Around 2000 different types of mutation in the *CFTR* gene have been described. One, a three-base deletion in the gene's DNA sequence that results in the loss of a single phenylalanine (represented as 'F') at position 508 of the protein (F508del or ΔF508), is the single most common mutation (see Figures 1.24–1.26, pages 27, 28). This mutation accounts for 70% of the underactive *CFTR* alleles in northern Europe and the US but is less common in southern Europe (45–55%), African

Americans (37%) and Ashkenazi Jews (30–35%). Most of the other mutations are individually uncommon, with occasional exceptions in association with particular ethnic groups (such as W1282X, which accounts for approximately 45% of mutant *CFTR* alleles in Ashkenazi Jews).

This wide variety of *CFTR* mutations means that it is not practicable to exclude all known mutations in screening for cystic fibrosis carrier status. In practice, a panel of common mutations is tested, which identifies approximately 90% of *CFTR* mutations in northern European populations. In antenatal screening programmes (where performed), mouthwash or blood samples for DNA analysis are collected from the pregnant woman and her partner. Her sample is tested with the panel of mutations and if it is found to be positive his sample is then tested. If his sample is also positive the pregnancy has a one in four risk of being affected and the couple can opt for prenatal diagnosis by fetal DNA analysis. This type of screening does not test for all known cystic fibrosis mutations and thus there is a residual risk even if either partner tests negative.

Carriers may also be identified following neonatal screening for cases of cystic fibrosis. In the UK, widespread neonatal screening for cystic fibrosis has been introduced. This is undertaken by initial measurement of the IRT on a blood spot. This protein is produced by the pancreas and its level in the blood is increased in cystic fibrosis, possibly as a result of abnormal pancreatic duct secretions. Although the sensitivity of the test is high, its specificity is not. Consequently, additional tests (including repeated IRT testing, DNA analysis and sweat testing) are undertaken if the IRT is elevated, in order to confirm the diagnosis. The detection of cystic fibrosis in children at an early stage has been found to be beneficial for their nutrition (see Southern et al. 2009 in Further reading) and possibly also for their long-term respiratory function (although this remains to be confirmed). In addition, of course, it permits the parents (who will both then be carriers) to be advised about the risk (one in four, or 25%) of recurrence of the condition in the next child that they conceive together and they can be offered prenatal diagnosis.

When either or both parents are found to be carriers of cystic fibrosis, there is also a need to consider extended family testing for carrier status.

## Family history

### DOWN SYNDROME

Most (95%) cases of Down syndrome are caused by an extra free copy of chromosome 21, resulting in 47 chromosomes in total, with three copies of chromosome 21 (trisomy 21, see Figure 1.3, page 7). In this situation, the parental karyotypes would be expected to be normal and the risk of recurrence for the parents of trisomy 21 or other major chromosomal abnormality at amniocentesis is 1.5% (risk at birth 1%). For mothers who are aged 40 years or more, the mother's age-related risk is used instead (Table 2.1, page 50). Thus, for example, the recurrence risk at birth (for trisomy 21) for a 43-year-old mother who has a previous child with trisomy 21 is 2% (one in 50).

Hence, for parents with a previous child or pregnancy affected by trisomy 21, there is a recurrence risk which is higher than the population risk, and many couples will seek reassurance by fetal DNA and/or chromosome analysis in future pregnancies. In this situation, the risks are not increased for other relatives above the general population risks and no special testing is indicated for them either before or during pregnancies. These relatives would not require blood karyotyping and would be offered routine antenatal screening tests for Down syndrome.

An important minority (5%) of cases of Down syndrome are, however, caused by chromosome translocations involving chromosome 21. The clinical features of translocation Down syndrome are indistinguishable from those due to trisomy 21 and karyotyping is required to establish the diagnosis. The karyotype of a patient with translocation Down syndrome is shown in Figure 2.14. This patient has 46 chromosomes in total, with two free copies of chromosome 21 and one extra copy of 21, which is joined to the top of chromosome 14. This translocation between chromosomes 14 and 21 might have occurred as a new event at the time the egg or sperm was produced or it might have been inherited. Hence, the parental karyotypes of a child or pregnancy with translocation Down syndrome must be examined. In this family, the mother's karyotype was normal but the father was found to have the translocation (Figure 2.15). He has only 45 chromosomes in total rather than the normal 46, as two free copies of 14 and 21 are replaced by a fused single chromosome. In this translocation, he has not lost or gained vital genetic material and hence he has no clinical features and is a balanced translocation carrier. The problem occurs when he comes to pass on his chromosomes to a pregnancy. Each sperm must receive one of each pair of chromosomes and this can result in sperm with normal chromosomes, balanced translocation chromosomes or unbalanced arrangements with gain or loss of chromosomal material (Figure 2.16).

From Figure 2.16, it might be expected that two of every four pregnancies would be chromosomally unbalanced, but in practice with this translocation (and with many other translocations) there is some

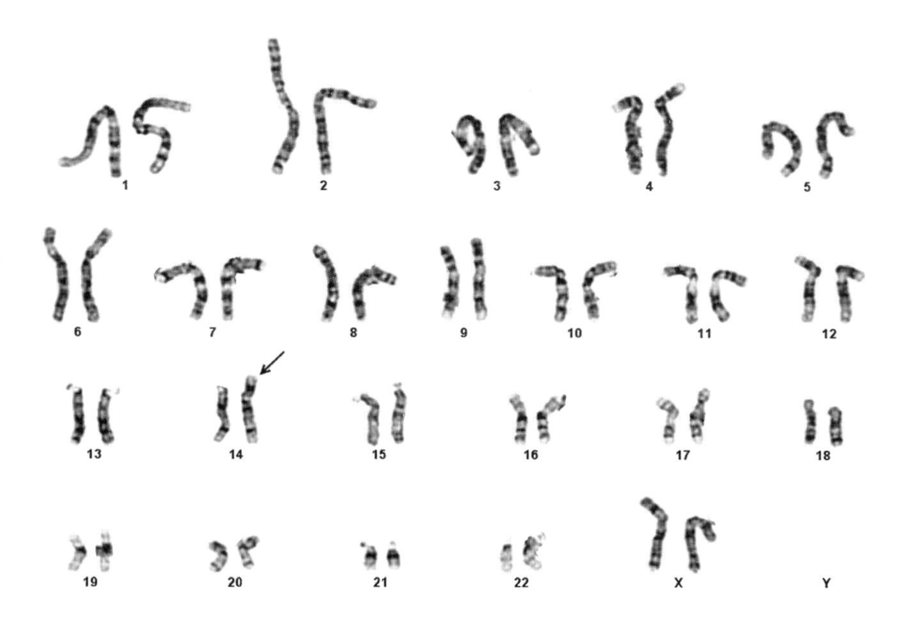


**Figure 2.14** Down syndrome attributable to an unbalanced translocation between chromosomes 14 and 21 (extra copy of chromosome 21 is arrowed)

selection against sperm (or eggs) with an imbalance. For example, the frequency of unbalanced arrangements for this particular translocation at the time of amniocentesis is 1% if the father is the carrier of the balanced translocation and 15% if the mother is the carrier. The couple needs to be offered fetal chromosome analysis after either CVS or amniocentesis in future pregnancies. The relative advantages and disadvantages of each will need to be discussed as outlined in the section on genetic causes of recurrent miscarriage (pp. 45–9).

In counselling the family with a chromosomal translocation, the clinical responsibility relates to the whole family. Thus, it is crucial to determine who else in the family also carries a balanced translocation. This can involve extended family studies and is usually undertaken by the regional genetics service.[1]

Translocations involving chromosome 21 can also involve other chromosomes and the risks to offspring depend not only on which chromosomes are involved but also on which parent is the carrier of the balanced translocation (Table 2.3).

In Figure 2.17, the woman is expecting her first child and, at booking, is noted to have a brother with Down syndrome. She does not know

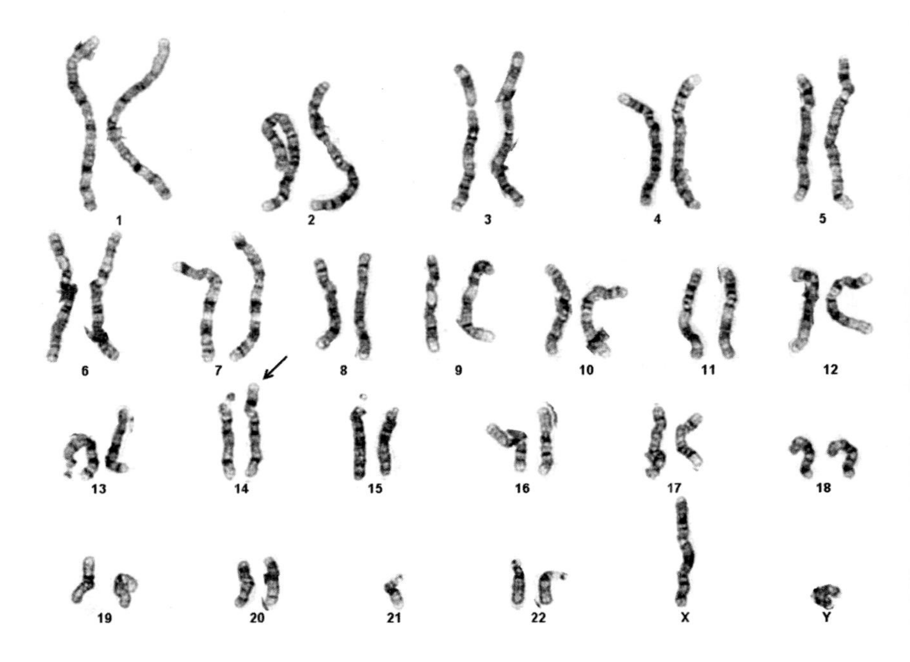

**Figure 2.15** Father with a balanced translocation between chromosomes 14 and 21 (arrowed)

whether he ever had a chromosome test and is anxious about an increased risk for Down syndrome in her pregnancy.

This is a common situation and the key is to try and discover whether or not her affected brother has trisomy 21 (in which case she can be reassured, given the general population risk and offered routine Down syndrome screening) or has translocation Down syndrome (in which case, further testing in the family is required).

The affected brother's full name, date of birth and addresses (at birth, early childhood and current) are helpful. The genetics centre local to the affected brother should be contacted.[1] The genetics centre should be able to confirm whether or not he has been tested and whether or not he has a translocation or trisomy 21. If no record of testing is found then it would be usual to offer the woman an urgent blood chromosome analysis to exclude a balanced translocation in her and the local genetics centre would organise testing on the affected brother and other relatives, as required. The alternative, of testing the brother first and then only testing the woman if a translocation is found, is usually impractical, given the timescale and potential timing of prenatal testing if she is found to be at risk.

**Table 2.3** Risks of chromosomally unbalanced offspring for carriers of balanced translocations involving chromosome 21

| *Translocation* | *Carrier* | *Risk of unbalanced arrangement at amniocentesis (%)* |
|---|---|---|
| 14–21 | Father | 1 |
| 14–21 | Mother | 15 |
| 21–22 | Father | 5 |
| 21–22 | Mother | 10–15 |
| 21–21 | Either parent | 100 |

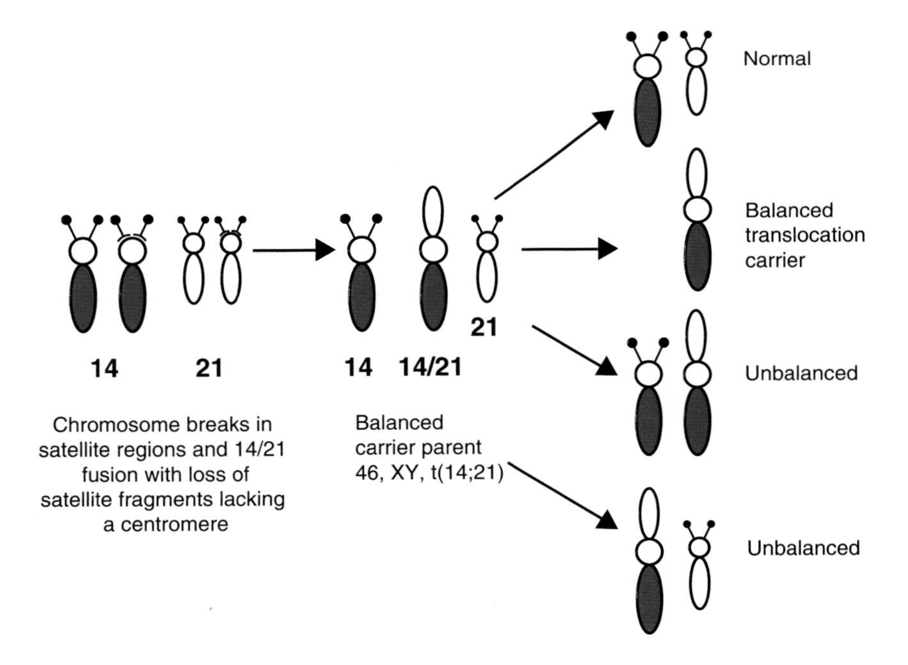


**Figure 2.16** Types of gamete for a carrier of a 14;21 chromosomal translocation

## NEURAL TUBE DEFECT

The neural groove appears at 20 days from conception; it is mainly closed by 23 days and fully closed by 28 days. Defective closure of the neural tube may occur at any level. Failure at the cephalic end produces anencephaly (Figure 2.11, page 56) or encephalocele (Figure 2.18), and failure lower down produces spina bifida (Figures 2.12 and 2.13, pages 57 and 58). Overall, anencephaly (with or without spina bifida)

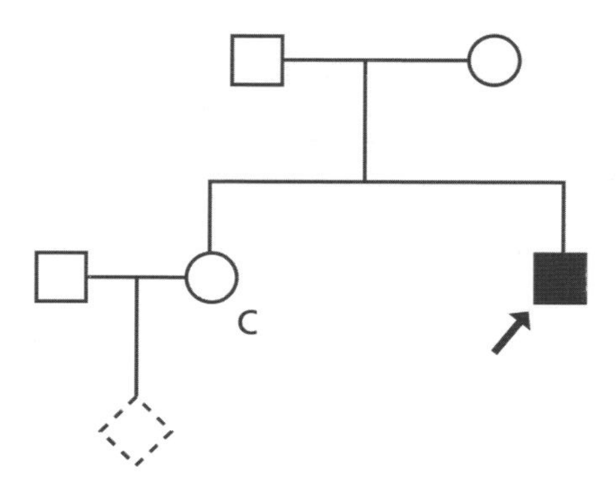


**Figure 2.17** Family history of Down syndrome

accounts for 40% of NTDs, spina bifida alone for 55% and encephalocele for 5%. Other malformations, particularly exomphalos and renal malformations, coexist in 25%.

The frequency of NTDs shows geographical variation and gradual change over time. In the US, Africa and Mongolia, one in 1000 births are affected. In South East England in the 1970s, three in 1000 births were affected and at that time the frequency was even higher (five to eight in 1000) in Ireland, Wales and the West of Scotland. Subsequently, the frequency has fallen throughout the UK and the figure (in 2009; see Scottish Perinatal and Infant Mortality and Morbidity Report 2010 in Further reading) in Scotland was 0.5 in 1000 births (1.02 if terminations are included). The reasons for this gradual fall are unknown but dietary improvements are believed to be a contributory factor.

Family and twin studies support multifactorial (part-genetic) inheritance for NTDs. The genetic components of susceptibility are unknown but one environmental component, maternal levels of folic acid during early pregnancy, has been identified.

In the UK, the recurrence risk of any NTD after an affected pregnancy is one in 25–33 (3–4%). The risk to the offspring of an affected woman is also one in 25–33 (3–4%). This risk is reduced to one in 100 by folic acid supplementation, which must be introduced prior to conception at a dosage (in this situation) of 5 mg/day. Prenatal diagnosis can be offered by means of detailed ultrasound scanning. If a couple has two or more affected children, the recurrence risk rises to one in ten.

For second-degree relatives (aunts/uncles, nephews/nieces) the risk is one in 70 and for third-degree relatives (cousins) it is one in 150.

**Figure 2.18** Encephalocele

## CYSTIC FIBROSIS

As described above, cystic fibrosis is an important cause of chronic morbidity and mortality in childhood. It is inherited as an autosomal recessive disorder and one in 22 of northern European and US populations are carriers. When two carriers have children, there is on average a one in four (25%) chance for each child that he or she will inherit both copies of the underactive *CFTR* gene and be affected.

As mentioned above, a wide variety of mutations (approximately 2000) in the *CFTR* gene have been identified. A 3 bp deletion, causing loss of one

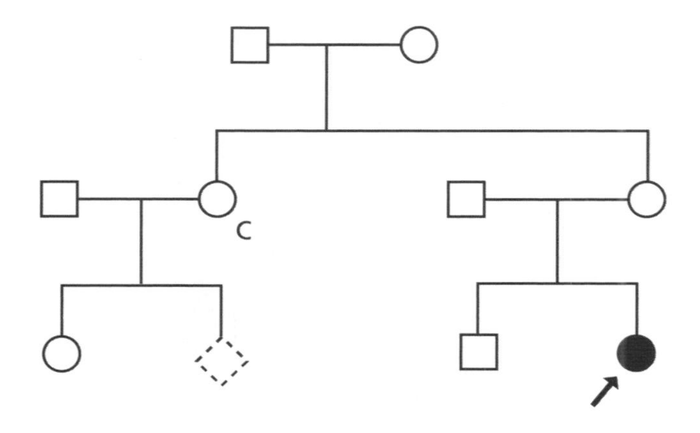


**Figure 2.19** Family history of cystic fibrosis

amino acid (phenylalanine or 'F') at amino acid position 508 (F508del or ΔF508), is the single most common mutation (see Figures 1.24–1.26, pages 27, 28). This mutation accounts for 70–80% of mutant cystic fibrosis gene copies in northern Europe and the US. Most other mutations are individually uncommon. About 50% of northern European and US patients have two copies of ΔF508 (ΔF508/ΔF508); most of the rest will have one copy of ΔF508 and one other mutant gene (ΔF508/M) and a minority will have two copies of other mutant (M) genes (M/M).

Figure 2.19 shows a family with cystic fibrosis. The woman is expecting her second child and was alarmed to hear that her niece has been diagnosed with cystic fibrosis. Her sister (the mother of the affected girl) must be a carrier of cystic fibrosis, as must her partner. DNA testing confirmed this and also identified that the pregnant woman's mother was a carrier (normal/mutant, N/M). The pregnant woman therefore had a one in two risk of being a carrier and her DNA test confirmed that she had inherited the mutant gene. Her husband is of North European origin and has no family history of the disease and so has the one in 22 general population carrier risk. Before he is tested, the risk to the pregnancy of being affected by cystic fibrosis is one in 88. This represents the pregnant woman's chance of passing the mutant allele on (1 x 0.5, i.e. her carrier risk multiplied by 0.5) combined with her husband's chance of passing on the mutant allele (one in 22 x 0.5, i.e. his carrier risk multiplied by 0.5). If testing identifies that he is a carrier, this risk to the pregnancy becomes one in four. If his screen for common cystic fibrosis mutations is negative, the risk to the pregnancy will be substantially reduced. If the cystic fibrosis screen detects 90% of mutations, the residual risk of the disease for the pregnancy will be approximately one in 880 – her carrier risk multiplied by 0.5 multiplied by his residual carrier risk multiplied by 0.5, i.e. 1 x 0.5 x (one in 22) x (one in ten) x 0.5 (although calculations involving Bayes' theorem can be used for greater accuracy in such calculations).

## MUSCULAR DYSTROPHY

A family history of muscular dystrophy always needs to be taken seriously. There are many subtypes of muscular dystrophy and most are inherited as single-gene disorders. This means that there are likely to be high recurrence risks for close family members. Moreover, for X-linked forms, even quite distant female relatives may still be at high risk. In light of the severity of many of these conditions, parents will commonly request reassurance by means of prenatal diagnosis.

### Duchenne muscular dystrophy

Duchenne muscular dystrophy (DMD) is the most common form of childhood muscular dystrophy, with a birth incidence of one in 3000 males. The gene for DMD is located on the short arm of the X chromosome and so males with a pathogenic mutation are always affected, whereas females with a single mutant *DMD* gene on one X chromosome and a normal copy on the other X chromosome are usually unaffected. Affected males are normal at birth but already have high levels of muscle enzymes such as creatine phosphokinase (CPK) in their serum. Walking may be delayed and they are rarely able to run properly. Muscle weakness is progressive and most will be wheelchair-bound by 13 years of age. Death from respiratory complications usually occurs at around 25 years of age.

Examination of the *DMD* gene in affected males usually (65%) shows a deletion of variable size; less commonly, a partial duplication or a point mutation is found. These changes, if found, can then be used to determine which females in the family are carriers and can also be used for the carriers to offer prenatal diagnosis. If the *DMD* mutation is not known in the family, carrier risks can be calculated from the mother's position in the pedigree, information from the mother's CPK levels, which are elevated in two-thirds of carriers, and by tracking the at-risk gene copy within the family using DNA markers.

In Figure 2.20, the woman marked "C" (for consultand) has a family history of DMD. Her brother died of DMD and her sister's son has recently been diagnosed with DMD. The mothers of each affected male must be carriers of the *DMD* mutation, as the alternative explanation of two new mutations is highly unlikely. The pregnant woman's mother is

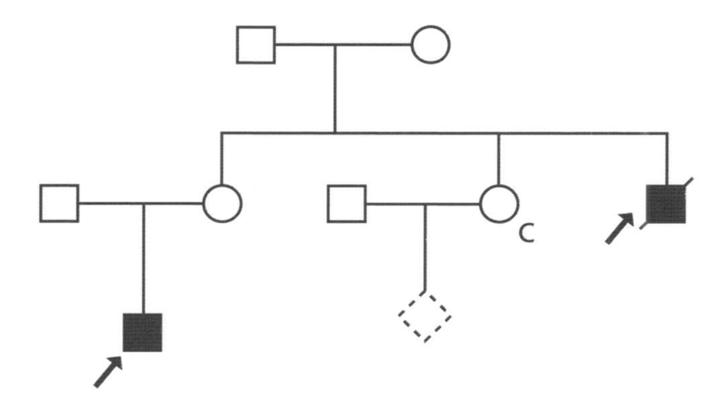


**Figure 2.20** Family history of Duchenne muscular dystrophy

therefore a carrier and must pass on either her X chromosome with the normal gene or the X chromosome with the *DMD* mutation. The pregnant woman's risk of being a *DMD* mutation carrier is thus one in two. The risk that her pregnancy is affected with DMD is one in eight (her carrier risk multiplied by her chance of passing on the mutant gene multiplied by the chance of a male pregnancy, i.e. 1/2 x 1/2 x 1/2). If the mutation is known and she is shown to be a carrier, this risk rises to one in four and fetal DNA analysis can be used to determine the sex of the fetus and the presence or absence of the *DMD* mutation. Usually, CVS is the source of the fetal DNA, as this allows the option of a first-trimester termination of pregnancy.

If the mutation in the affected males has not been identified, carrier detection and prenatal diagnosis may still be possible using other methods but will lack 100% accuracy. In this situation, many couples still choose to terminate an at-risk pregnancy because of the severity of the disorder.

Carrier detection within the extended family will be required and every family with DMD should be referred to the regional genetics centre.[1]

## Myotonic dystrophy

Myotonic dystrophy (dystrophia myotonica) is the most common adult onset form of muscular dystrophy, with a frequency of one in 9000. The gene (*DMPK*) for myotonic dystrophy is located on chromosome 19 and the condition is inherited as an autosomal dominant trait. Thus, each affected person has one normal *DMPK* allele and one mutant

*DMPK* allele. Unlike most genetic mutations, the *DMPK* mutation is unstable and varies in size in individuals, even in the same family (see Figure 1.23, page 26). The mutation tends to increase in size with each generation, particularly when transmitted by the mother. There is an approximate correlation between the size of the mutation and the severity of the condition, with larger mutations causing earlier onset and more severe disability. At the mildest end of the spectrum, there may be few or no symptoms of muscle weakness and perhaps only adult-onset cataracts. More commonly, the affected person has adult-onset progressive muscle weakness, especially of the face, sternomastoids and distal limb muscles. Severe disability is usually 15–20 years from the onset. General anaesthesia may be hazardous and the anaesthetist should be warned of the patient's diagnosis even if there are few or no symptoms. At the most severe end of the spectrum the fetus can be affected. In this situation, almost invariably, the mother is the affected parent and the fetus has a *DMPK* allele that is massively expanded. Neonatal hypotonia is marked and these children later show major learning difficulties and progressive muscle weakness. In each of these situations, the clinical diagnosis can be confirmed by DNA analysis. Within a family, presymptomatic testing of healthy at-risk adults and prenatal diagnosis are also possible by DNA analysis.

A significant number of families only come to light following the birth of a severely affected infant. Earlier identification may be possible at the antenatal clinic by recognition of the typical myotonic dystrophy facies with a long thin, expressionless face and ptosis. Sufferers may also demonstrate myotonia, with delayed relaxation after muscle contraction (for instance, after shaking hands).

There are a number of obstetric complications in women with myotonic dystrophy. There may be delay in labour, requiring intervention, and great care must be exercised in using appropriate anaesthetic agents. Postpartum haemorrhage is also common.

There are general medical complications in myotonic dystrophy. These include cardiac rhythm abnormalities and diabetes mellitus, which should be excluded during pregnancy and monitored afterwards.

A pregnant woman with myotonic dystrophy has a one in two risk of transmitting the disorder. Counselling about prenatal diagnosis is difficult because an affected pregnancy may result in either a severely affected neonate or an infant with childhood or adult onset of symptoms. DNA analysis cannot completely distinguish these outcomes. Delivery of an at-risk child should thus take place in an obstetric unit with an attached neonatal intensive care unit.

In every family, genetic counselling of the extended family will be required and they should be referred to the regional genetics centre.[1]

## LEARNING DIFFICULTIES

Moderate and severe learning difficulties (IQ of less than 50) affect 1% of newborns but this figure falls to 0.3–0.4% in children of school age, owing to deaths in infancy from associated abnormalities or rapidly progressive disorders. The cause can be identified in about 75% of these children (Table 2.4).

Trisomy 21 is the single most common cause of major learning difficulties in this age group; a variety of other chromosome abnormalities account for another 2%. Over 250 single-gene disorders have been described for which major learning difficulty is a consistent or common feature. Numerically, the most frequent among the autosomal dominant disorders is tuberous sclerosis. Among autosomal recessive disorders, phenylketonuria, cerebral degenerative disorders and recessive microcephaly predominate, whereas among the X-linked disorders the fragile X syndrome is the single most common cause.

Genetic assessment will thus be directed towards establishing the cause of the disability in the family and referral to the regional genetics centre will be required.[1] When the cause is obscure and no other relatives are affected to suggest a pattern of inheritance, observed (empiric) recurrence risks from the outcome of a large number of similar families need to be used. In the absence of a

**Table 2.4 Causes of moderate and severe learning difficulties in children of school age**

| *Cause* | *Occurrence (approximate %)* |
|---|---|
| **Chromosomal disorders** | |
| Trisomy 21 | 25 |
| Other | 2 |
| **Single-gene disorders** | |
| Autosomal dominant | 1 |
| Autosomal recessive | 10 |
| **X-linked** | |
| Fragile X syndrome | 4 |
| Other | 4 |
| **Other** | |
| Brain malformations/dysmorphic syndromes* | 14 |
| Environmental factors | 15 |
| Unexplained | 25 |

* Excluding recognised chromosomal and single-gene disorders.

specific diagnosis, reassurance by prenatal diagnosis during a subsequent pregnancy cannot be offered.

## Fragile X syndrome

The fragile X syndrome is caused by an unstable length mutation in the *FMR1* gene on the X chromosome. Relatively small increases in length (so-called premutations) of this gene do not cause learning difficulties, although they have been reported to cause premature ovarian failure in 20% of women carrying them and fragile X tremor/ataxia syndrome (FXTAS) in 40% of males possessing such premutations. Once these mutations have expanded beyond a critical size, however, each male possessing one of them will be affected by learning difficulties. Females may carry a similar pathogenic mutation but are usually, although not always, protected by the presence of a normal-sized gene on the opposite X chromosome. The diagnosis in affected males can be confirmed by DNA analysis and this test can also detect the carriers and be used for prenatal diagnosis.

In Figure 2.21, the consultand has two male cousins with fragile X syndrome. DNA testing revealed that their mother was a carrier and that she had inherited the mutation from her father. He was healthy but had a small premutation (and is therefore termed a 'normal transmitting male'). As he had to pass on his X chromosome to each daughter, the consultand's mother must be a carrier and the consultand has a one in two risk of being a carrier. The risk to her pregnancy (if her

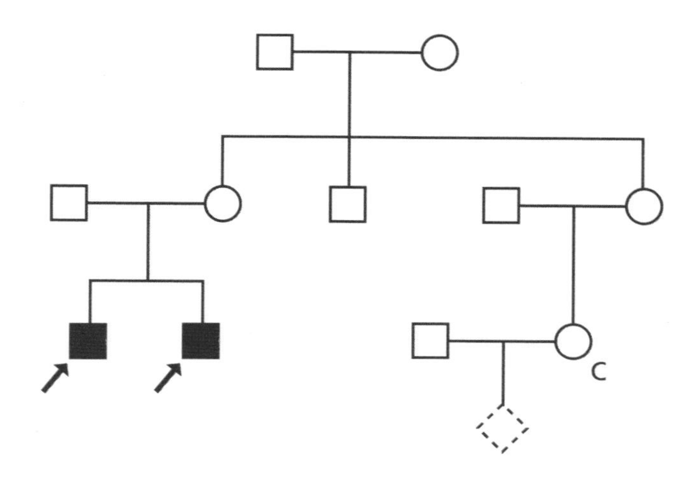


**Figure 2.21** Family history of fragile X syndrome

mother has a full fragile X mutation) is thus one in eight (her carrier risk multiplied by her chance of passing on the mutant gene multiplied by the chance of a male pregnancy, i.e. 1/2 x 1/2 x 1/2). If she is shown to be a carrier, this risk rises to one in four and fetal DNA analysis can be used to determine the sex of the fetus and the presence of the *FMR1* mutation. Usually CVS is the source of the fetal DNA, as this allows the option of a first-trimester termination of pregnancy.

Carrier detection within the extended family will be required and every family with fragile X syndrome should be referred to the regional genetics centre.[1]

## CONGENITAL MALFORMATIONS

Three percent of newborns have a single major congenital malformation and 0.7% of newborns have multiple major malformations. Minor congenital malformations (for example, a single umbilical artery, soft-tissue syndactyly or ear tags or pits) are even more common but, if multiple, should alert the clinician to the possibility of an associated major malformation. Table 2.5 indicates the identifiable causes of major congenital malformations.

As indicated, the basis of most congenital malformations is unknown. Multifactorial (part-genetic) inheritance is the most common identifiable cause followed by single-gene disorders and chromosomal disorders. Thus, genetic conditions account for at least one-third of all congenital malformations.

Visible duplication or deletion affecting any of the autosomes is almost invariably associated with significant learning difficulties, postnatal growth deficiency and an unusual facial appearance. Multiple congenital malformations and intrauterine growth restriction are also commonly seen and roughly correlate in severity with the extent of the

**Table 2.5 Aetiology of major congenital malformations**

| *Cause* | *Occurrence (approximate %)* |
|---|---|
| Idiopathic | 50.0 |
| Multifactorial (part-genetic) disorders | 30.0 |
| Single-gene disorders | 7.5 |
| Chromosomal disorders | 6.0 |
| Maternal illness | 3.0 |
| Congenital infection | 2.0 |
| Drugs, X-rays, alcohol | 1.5 |

chromosomal imbalance. Over 250 single-gene disorders include major congenital malformations as a consistent or frequent feature. Recognition of these single-gene disorders and of inherited structural chromosome rearrangements is of clinical importance in view of their potential high recurrence risks.

Maternal illnesses associated with an increased risk of fetal malformation include type 1 (insulin-dependent) diabetes mellitus, epilepsy, alcohol misuse and phenylketonuria. There is a 5–15% risk of congenital malformation (especially congenital heart disease, NTD and sacral agenesis) for the offspring of a mother with diabetes, in inverse proportion to the quality of her diabetic control. The risk is also increased (to about 6%, especially for cleft lip and congenital heart disease) for a mother with epilepsy, although here it is difficult to separate the risk attributable to the disease and that attributable to her medication. Untreated maternal phenylketonuria carries a high risk (25%) to the fetus for significant learning difficulties, microcephaly and congenital heart disease.

Genetic counselling in families with a family history of congenital malformations will thus depend heavily on the identification of the underlying cause. In the absence of an identified cause, observed (empiric) recurrence risks derived from the outcome of large numbers of similar families, are used. Prenatal diagnosis may be possible using high-resolution ultrasound scanning.

## OVARIAN CANCER

Ovarian cancer affects approximately one in 54 women in the UK and familial ovarian cancer accounts for 5–10% of the total. The familial form should be suspected particularly if the affected woman is young (under 40 years of age), has bilateral disease, has associated tumours or has a relevant family history of ovarian or breast cancer. Specific referral criteria are usually available from the local medical genetics centre (e.g. www.nhsggc.org.uk/content/default.asp?page=s1154_3). Noninherited ovarian cancer has a mean age of onset of 70 years whereas the inherited form often occurs in the 40s and 50s. In noninherited cancer, mutations accumulate in a clone of cells, which then undergoes malignant transformation. As none of these mutations was inherited, comparable changes are unlikely in the other ovary. By contrast, in the inherited form, the first key genetic mutation is inherited and is present in all cells. This accounts for the earlier age of onset and increased chance of bilateral involvement in the inherited form.

One of the most common causes of inherited ovarian cancer is familial breast/ovarian cancer. This is inherited as an autosomal dominant trait and in many cases is due to mutations in the *BRCA1* gene on the long arm of chromosome 17. According to meta-analyses published by Antoniou et al. (see Further reading), for females with a *BRCA1* mutation, the risk by age 70 for breast cancer is approximately 65% with a 95% confidence interval (CI) of 44–78%, and for ovarian cancer is 39% (95% CI 18–54%). Thus, within the family, some women with the mutant gene will have only breast cancer, some only ovarian cancer, some will have both and some will have neither. Males with the mutant gene usually do not develop cancer but will, on average, transmit the mutant gene to 50% of their offspring. Thus, a female related to other affected females through an unaffected male is still at significant risk.

Less commonly, breast/ovarian cancer results from mutations in the *BRCA2* gene on chromosome 13, which in females cause risk by age 70 of breast cancer of approximately 45% (CI 31–56%) according to a meta-analysis by Antoniou et al. (2003). The risk of ovarian cancer was reported in 2003 to be 11% (CI 2.4–19%) although a subsequent meta-analysis found it to be up to 29% (Antoniou et al. 2009). Ovarian cancer can also result from pathogenic mutations in the *MLH1* or *MSH2* genes, which more commonly cause colorectal or endometrial cancer as part of the inherited syndrome known as hereditary nonpolyposis colorectal cancer (HNPCC), mentioned below.

In Figure 2.22, the consultand is concerned about her family history of cancer. Multiple women have had variable combinations of breast and/or ovarian cancer and this pattern would be highly suspicious of familial breast/ovarian cancer. DNA analysis of the *BRCA1* and *BRCA2* genes in an affected woman may confirm that this is the case. If a causative mutation is identified in an affected relative, the consultand

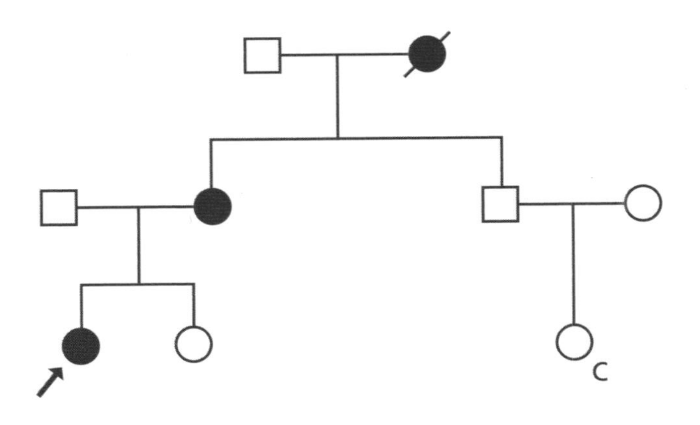


**Figure 2.22** Family history of cancer

can then be offered presymptomatic genetic testing and, if this is found to be positive, she can be offered appropriate clinical screening. Counselling of other family members at risk will be required and the family should be referred to the regional genetics centre (see Directory of UK Genetics Centres in Appendix 1).

### UTERINE CANCER

Endometrial cancer is one of the most common types of tumour that result from mutations in one of the genes (especially *MLH1* and *MSH2*), which encode proteins that are involved in the physiologically important DNA 'mismatch repair' process. Such mutations cause the autosomal dominantly inherited tumour predisposition syndrome HNPCC, mentioned above. Those who possess such a mutation have high risks of colorectal cancer, necessitating regular screening colonoscopies, and of endometrial cancer (which affects 40–60% of women possessing such a mutation). In addition, there is a smaller increased risk of upper gastrointestinal tumours, for which appropriate screening endoscopies may also be undertaken.

## Reference

1. British Society for Genetic Medicine. Genetics Centres (UK): www.bsgm.org.uk/information-education/genetics-centres

# Section Three

## Clinical case scenarios

# Clinical case scenarios

## Introduction

This section applies the knowledge gained in the previous two sections to clinical situations.

## Case 1: Unexpected finding at amniocentesis

### SITUATION

An expectant mother underwent amniocentesis because of an elevated screening risk for Down syndrome and the cytogenetic laboratory has noted an unusual appearance of a chromosome (Figure 3.1). Which chromosome is causing concern? What is the significance of this finding?

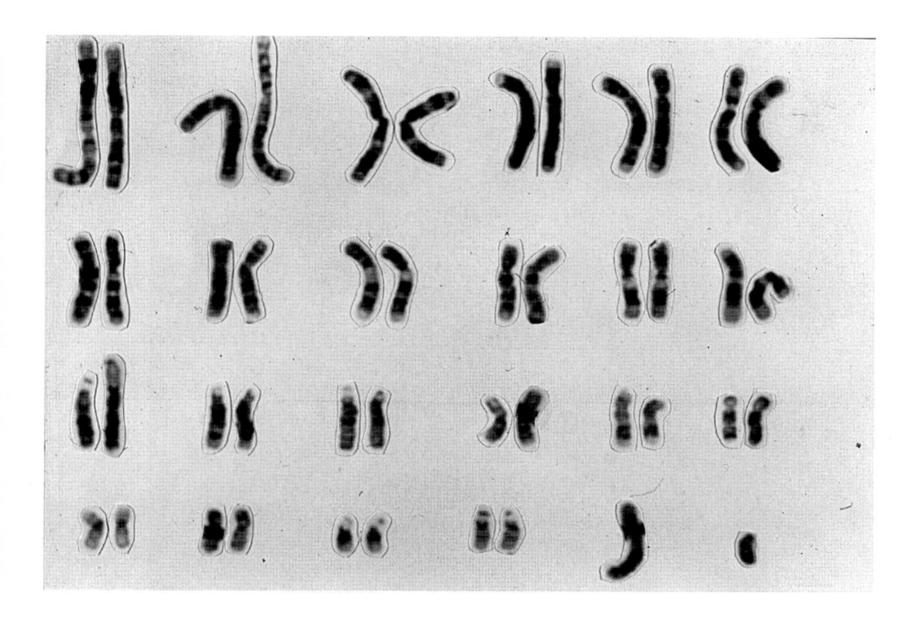

**Figure 3.1** Karyotype for interpretation

## CLINICAL RESPONSE

The short arm of chromosome 13 is longer than normal. The short arms of chromosomes 13–15, 21 and 22 are quite variable in length and reflect the number of copies of duplicated ribosomal genes in these areas. This is thus suspected to be a chromosomal variant or polymorphism of no clinical significance. This can be confirmed by checking the parents' blood karyotypes and finding that a healthy parent has the same chromosomal variant.

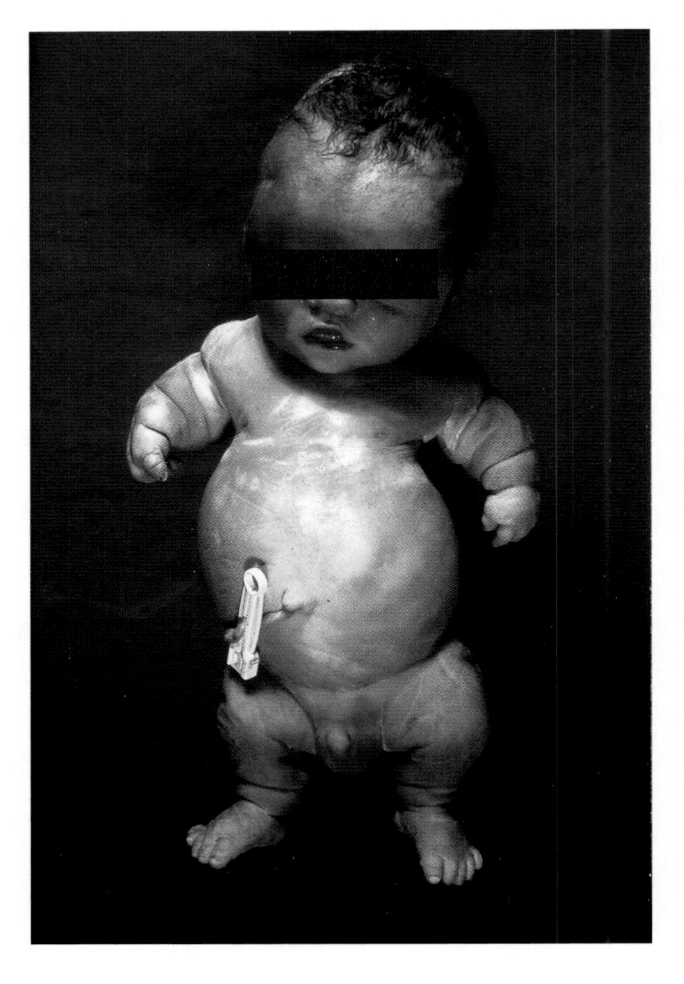

**Figure 3.2** Lethal short-limbed skeletal dysplasia

## Case 2: Lethal short-limbed skeletal dysplasia

### SITUATION

Polyhydramnios in an otherwise uneventful pregnancy led to detailed ultrasound scanning. This revealed marked shortening of the limbs and the child was stillborn (Figure 3.2). What must the clinician do in this situation?

### CLINICAL RESPONSE

There are multiple causes of lethal skeletal dysplasia. Some have high recurrence risks and some have low recurrence risks. A full body X-ray is crucial for distinguishing these different types and a DNA sample can also be extremely useful in confirming the suspected diagnosis. DNA can be extracted at postmortem from cardiac blood, a sample of liver or spleen, or from cultured skin fibroblasts. Skin fibroblasts can still be grown from a skin biopsy as late as three to four days after death. Ideally, a full post-mortem examination should be performed but, if the parents refuse consent, they will often allow an X-ray and a cardiac blood sample. For couples with a high risk of recurrence, prenatal detection can be offered by means of serial detailed ultrasound scans to monitor long-bone growth. If a pathogenic DNA abnormality has been identified in the stillborn infant, then prenatal diagnosis by DNA testing should also be possible.

## Case 3: Family history of Down syndrome

### SITUATION

In the obstetric clinic, a primigravida at 10 weeks of gestation reports that her cousin in the US has just had a baby with Down syndrome. What action is required?

### CLINICAL RESPONSE

Clinically, it is impossible to distinguish the more common low-risk trisomy 21 from the less common, potentially high-risk translocation Down syndrome. Full details of the affected child (name, date of birth, mother's name and address) should be taken and the regional genetics centre will contact the US centre to see if chromosomal analysis has been performed. If it has and the result is trisomy 21, our patient can be reassured that she has only the general population risk and offered standard first- or second-trimester Down syndrome screening. If the affected child has translocation Down syndrome then our patient needs an urgent blood chromosome analysis to ensure that she is not a translocation carrier. If she were, the couple would need to be counselled and offered prenatal diagnosis. If contact with the US centre is not possible then, despite the low risk, blood chromosome analysis should be performed on our patient to exclude a translocation.

## Case 4: Family history of Huntington disease

### SITUATION

An elderly primigravida (38 years of age) is concerned about her family history of Huntington disease, which affects her father and affected her paternal grandmother (now deceased) (Figure 3.3). Using the web-based resource Online Mendelian Inheritance in Man (OMIM) (Appendix 1), is it possible to determine the mode of inheritance of Huntington disease? What is her risk at 38 years of age? What is the risk to the fetus? Are DNA-based tests available to help clarify these risks?

### CLINICAL RESPONSE

Huntington disease is inherited as an autosomal dominant trait with age-dependent expression. The gene is located near the tip of the

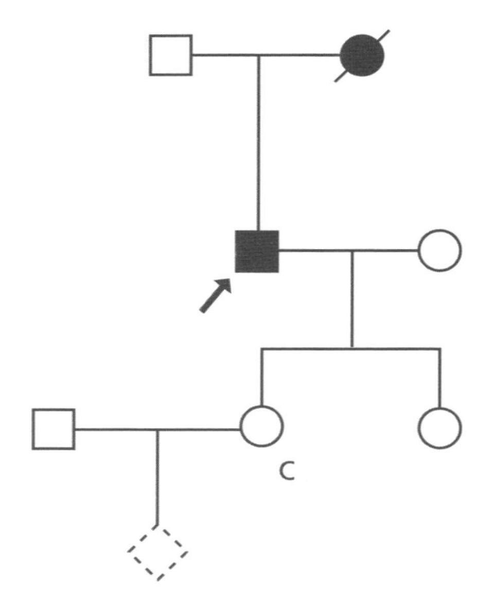


**Figure 3.3** Family history of Huntington disease

short arm of chromosome 4 and each affected person thus has one mutant gene and a normal gene on the opposite copy of chromosome 4. The risk to our patient that she has inherited the mutant gene from her father is one in two. The condition may present in childhood but this is unusual and onset in the 40s or 50s is more common. Thus, although our patient is currently healthy, she is too young to significantly reduce her risk from one in two. The risk to her fetus is one-half of her own risk or one in four.

A DNA test for Huntington disease is available and affected patients show a length mutation with the mutant gene being larger in size than the normal gene. This length mutation is unstable (as is the case with myotonic dystrophy and fragile X syndrome) and may change in size on transmission, especially from a male. The couple needs careful counselling to ensure that they understand the implications of DNA testing and this will usually be undertaken by the regional genetics centre. If she does not carry the mutant gene for Huntington disease then the risk to the fetus is negligible and prenatal testing for this condition is not indicated.

## Case 5: Family history of Duchenne muscular dystrophy

### SITUATION

In his referral letter to the obstetric booking clinic, the general practitioner notes that the pregnant woman's brother died of DMD and so he has measured her level of serum creatine kinase and found it to be within normal limits. Can she be reassured?

### CLINICAL RESPONSE

This woman's pregnancy is at high risk of DMD, despite her normal level of serum creatine kinase, and she needs an urgent referral to the regional genetics centre to clarify the magnitude of this risk. DMD is inherited as an X-linked recessive trait and, in the absence of any other family history, the mother of an affected son has a two-thirds risk of being a carrier. If the mother's levels of serum creatine kinase are elevated, this will confirm that she is a carrier, although normal levels are found in one-third of known carriers. Without creatine kinase data, her daughter will have a one in three chance of being a carrier and thus will have a one in six chance that a pregnancy, if male, will be affected. As is the case for her mother, an elevated level of creatine kinase will raise the daughter's risk of being a carrier but a normal level will not exclude the chance that she is a carrier. This is further confounded by measuring levels of creatine kinase during pregnancy, when the levels are reduced in both carriers and noncarriers.

In the genetics clinic, information from the family tree and creatine kinase levels will be combined to give a risk to the pregnancy and this will be supplemented, when possible, by DNA analysis, which will be helpful for both precise carrier detection and prenatal diagnosis.

## Case 6: Unexplained high level of maternal serum alphafetoprotein (MSAFP)

### SITUATION

Routine antenatal screening reveals a high level of MSAFP but detailed ultrasound scans are normal.

### CLINICAL RESPONSE

MSAFP can be elevated for a variety of causes. One of the most common causes is an underestimated gestation. The ultrasound scans will exclude anencephaly, encephalocele and the majority of cases of spina bifida. They will also exclude a delayed miscarriage, anterior abdominal wall defects and teratoma. More rare identifiable causes include fetal skin defects, placental haemangioma, congenital nephrotic syndrome and maternal hereditary persistence of alpha fetoprotein.

Unexplained elevations of MSAFP are associated with an increased risk of spontaneous miscarriage, stillbirth, low birthweight and perinatal death and thus these pregnancies merit follow-up.

## Case 7: Family history of siblings with Goldenhar syndrome

### SITUATION

A colleague asks your advice about an Australian family with Goldenhar syndrome. This affects a brother and sister and their aunt is expecting her first child.

### CLINICAL RESPONSE

OMIM (Appendix 1) can be used to look at the genetics and clinical features of Goldenhar syndrome. This reveals that Goldenhar syndrome is usually not inherited and hence two affected persons in the same family would be most unusual. This would bring the diagnosis into question and urgent assessment by a clinical geneticist is required. (In this family, the children had learning disabilities and dysmorphic features caused by an inherited translocation.)

This family emphasises the need to be really secure about the diagnosis before proceeding with genetic counselling and prenatal diagnosis. Assessment of clinical syndromes can be especially difficult as only rarely does an affected person have every clinical feature. Terms such as atypical case or *forme fruste* are commonly associated with an incorrect diagnosis.

## Case 8: Family history of microcephaly

### SITUATION

The pedigree of a family with a child (Figure 3.4) with microcephaly is shown in Figure 3.5.

### CLINICAL RESPONSE

Microcephaly reflects defective brain growth and has multiple causes. In this family, the parents are blood relatives (first cousins) and this makes the diagnosis of autosomal recessive microcephaly highly likely. Confirmation of this by DNA testing is possible only in some cases. The couple would be counselled that this is the most likely diagnosis and that in this event the recurrence risk is one in four. Prenatal diagnosis by serial ultrasound measurements of fetal head growth could be offered in a future pregnancy but the reduced head growth may not be evident until late in the pregnancy.

Other causes of microcephaly include congenital infection (rubella, cytomegalovirus or toxoplasmosis), birth trauma, chromosomal imbalance, fetal alcohol syndrome and maternal phenylketonuria. The last of these needs to be considered in families with affected siblings with microcephaly before concluding that autosomal recessive microcephaly is the cause.

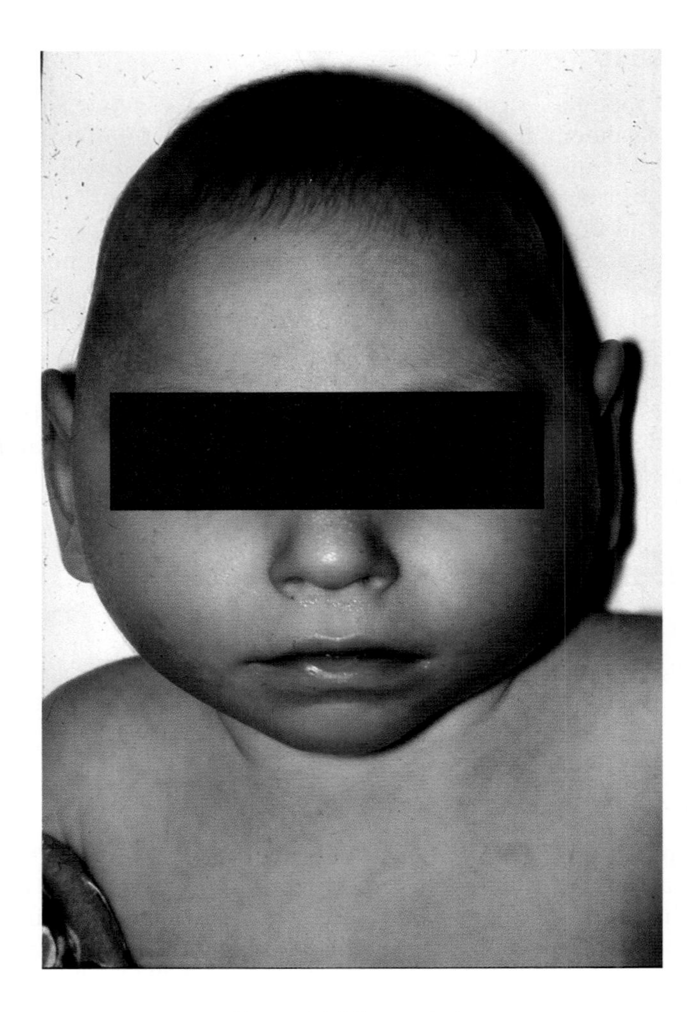

**Figure 3.4** Microcephaly

## Case 9: Unexpected finding at amniocentesis

### SITUATION

Amniocentesis was performed in view of an elevated maternal serum screening risk of Down syndrome and the karyotype revealed 47,XXY.

### CLINICAL RESPONSE

This is the karyotype of Klinefelter syndrome (see Figure 2.1, page 44). The parents will need careful counselling in order to make an

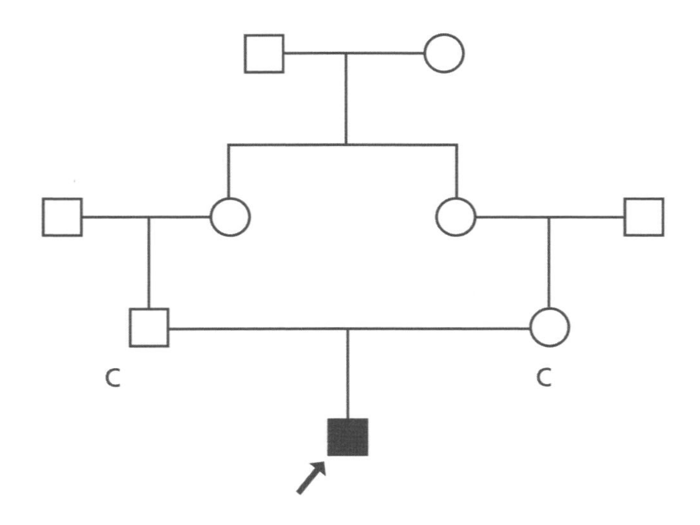


**Figure 3.5** Family history of microcephaly (affected child is shown in Figure 3.4)

informed decision about the pregnancy. The older medical textbooks gave an ill-informed picture and information from unselected newborn surveys reveals that most children with Klinefelter syndrome are clinically normal. There is a 10–20-point reduction in verbal skills but performance scores are usually normal and learning disabilities are uncommon. As an adult, without assisted conception, infertility occurs unless there is mosaicism, and testosterone-replacement therapy will be required from early adolescence. The overall birth incidence of Klinefelter syndrome is one in 1000 males and the recurrence risk for a family does not appear to be increased above this general population risk.

## Case 10: Family history of Down syndrome

### SITUATION

At booking, a mother reports that her brother has Down syndrome and she was told that she would need a test during any pregnancy.

### CLINICAL RESPONSE

The key is to determine whether this is the more common trisomy 21 type of Down syndrome or the less common chromosomal translocation form. If her brother has trisomy 21, her recurrence risk is not increased above the general population risk (see Table 2.1, page 50), whereas if he has a translocation her pregnancy may be at substantial risk.

The local regional genetics centre should be contacted with the full name, date of birth and address in early childhood for the affected brother. Their records will show if he has trisomy 21 or a translocation. If he has a translocation, family members might already have been tested. This was the case with this family and the sister was known to carry a translocation involving chromosome 21. In this particular case, there was a high risk (15%) of Down syndrome for her pregnancy and prenatal diagnosis by amniocentesis or CVS was offered. In this situation, offering only a first-trimester combined (ultrasound and biochemical screening) test to detect Down syndrome would be inappropriate, because of its lower sensitivity.

## Case 11: Importance of genetic ancestry

### SITUATION

At booking, a healthy Cypriot mother is found to have a hypochromic microcytic anaemia.

### CLINICAL RESPONSE

Most commonly, a hypochromic microcytic anaemia in pregnancy will be attributable to iron deficiency. Occasionally, there are other causes which might be suspected in this situation given the genetic ancestry of the mother. One in six Cypriots is a carrier for beta-thalassaemia. These carriers are healthy but have microcytosis (with a mean corpuscular volume less than 80 fl) and a low mean cell haemoglobin (mean corpuscular haemoglobin less than 27 pg/cell). In contrast to iron-deficiency anaemia, the iron stores are normal and haemoglobin $A_2$ is increased (greater than 3.5%).

If carrier status for beta-thalassaemia is confirmed, the woman's partner will also need to be tested. If both are carriers then the pregnancy has a one in four risk of beta-thalassaemia, which causes a severe chronic anaemia with a need for recurrent blood transfusions. Over 120 different mutations in the beta-globin gene cluster may be responsible but, in each population group, a subset of mutations is particularly common. DNA analysis can be used for prenatal diagnosis once the causative mutations in a family are known. Other family members also need to be offered genetic counselling and carrier testing.

Beta-thalassaemia carriers are particularly frequent in Mediterranean countries and South-East Asia. Other examples of conditions where the genetic ancestry is important are sickle cell disease, alpha-thalassaemia and Tay-Sachs disease.

Sickle cell disease is inherited as an autosomal recessive trait and is caused by a specific mutation in the beta-globin gene. It is especially prevalent in people of African or Caribbean descent, people from the Mediterranean, India and the Middle East. Routine haematological tests are normal in carriers and the haematology laboratory will perform specific tests if alerted by the clinician. If both parents are carriers there is a one in four risk that the fetus will have sickle cell disease. This is a severe chronic haemolytic anaemia with associated infarctions resulting from vascular obstruction. Prenatal diagnosis can be offered by DNA analysis.

Alpha-thalassaemia is inherited as an autosomal recessive trait and is caused by a variety of mutations in the alpha-globin gene cluster. The severe form of alpha-thalassaemia is caused by a complete lack of alpha-globin genes. In this case there is a profound in utero anaemia with hydrops fetalis and intrauterine or early neonatal death. The carrier parents of this severe form show a hypochromic microcytic anaemia, which needs to be distinguished from iron-deficiency anaemia and beta-thalassaemia carrier status. Carriers of the severe form of alpha-thalassaemia are particularly frequent in South-East Asia. Prenatal diagnosis for at-risk pregnancies can be offered by DNA analysis.

Tay-Sachs disease needs to be considered when the parents are Ashkenazi Jews. Tay-Sachs disease is inherited as an autosomal recessive trait and results in a progressive incurable neurodegeneration in childhood. Carrier parents are healthy but have reduced serum and leucocyte beta-*N*-acetylhexosaminidase A levels. Carrier screening and prenatal diagnosis can be offered by enzyme analysis or DNA analysis.

## Case 12: Never say never

### SITUATION

At a prepregnancy clinic, a prospective mother raises her concerns about a limb defect in her nephew (Figure 3.6).

### CLINICAL RESPONSE

The limb defect in this child is a typical transverse amputation defect. Most, if not all, are believed to be attributable to in utero amputations by strands of amnion produced by premature rupture of the amnion. If

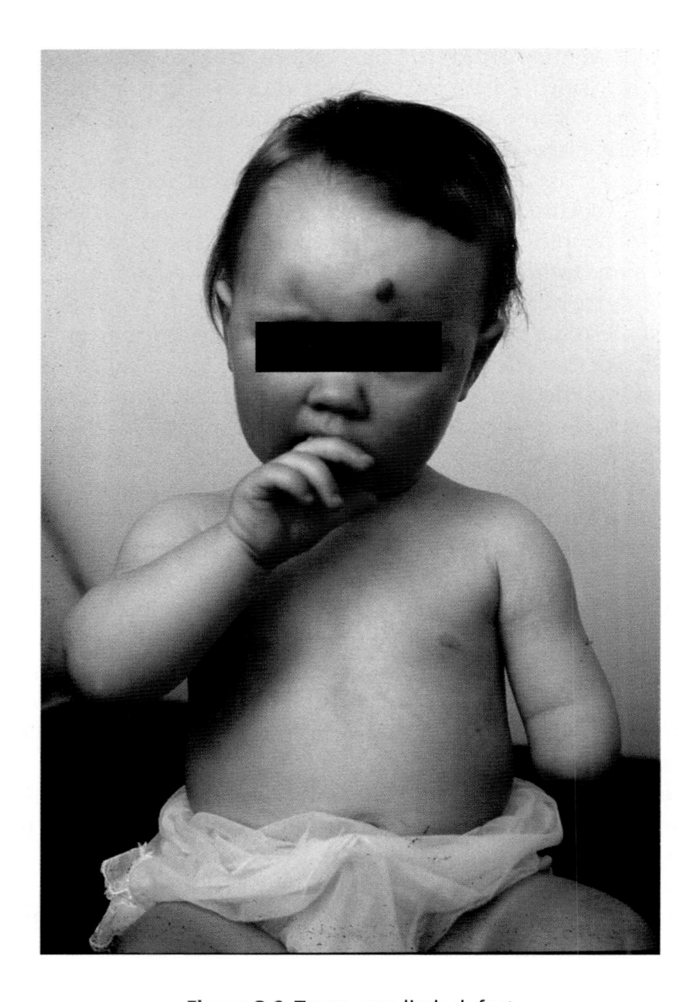

**Figure 3.6** Transverse limb defect

the chorion is also involved, there may be a history of leakage of liquor and the child may also show congenital deformations due to oligohydramnios. The amniotic bands may produce a variety of other defects, including asymmetric facial clefts.

The general population frequency is one in 5000 and in this situation the prospective mother would be reassured that her risk was low but not nonexistent (as we all have the general population risk). Further reassurance can be offered if required in a future pregnancy by detailed ultrasound scanning.

## Case 13: Unexpected finding at amniocentesis

### SITUATION

At amniocentesis, performed as a consequence of an increased maternal serum screening risk for Down syndrome, an apparently balanced chromosomal translocation between chromosomes 11 and 22 is detected (Figure 3.7). What action is indicated?

### CLINICAL RESPONSE

The cytogenetic report included the word 'apparently' as the lower limit of resolution of DNA loss or gain with the light microscope is 4 Mb. In the creation of a translocation, chromosomal breaks occur and DNA may be damaged or lost at the breakpoints. In order to exclude this possibility, the first step is to check the parental chromosomes.

If a healthy parent carries the same translocation then the couple can be reassured that the fetus is unlikely to be affected by the chromosomal change. The couple will need counselling in terms of the possibility of chromosomally unbalanced offspring in future pregnancies and

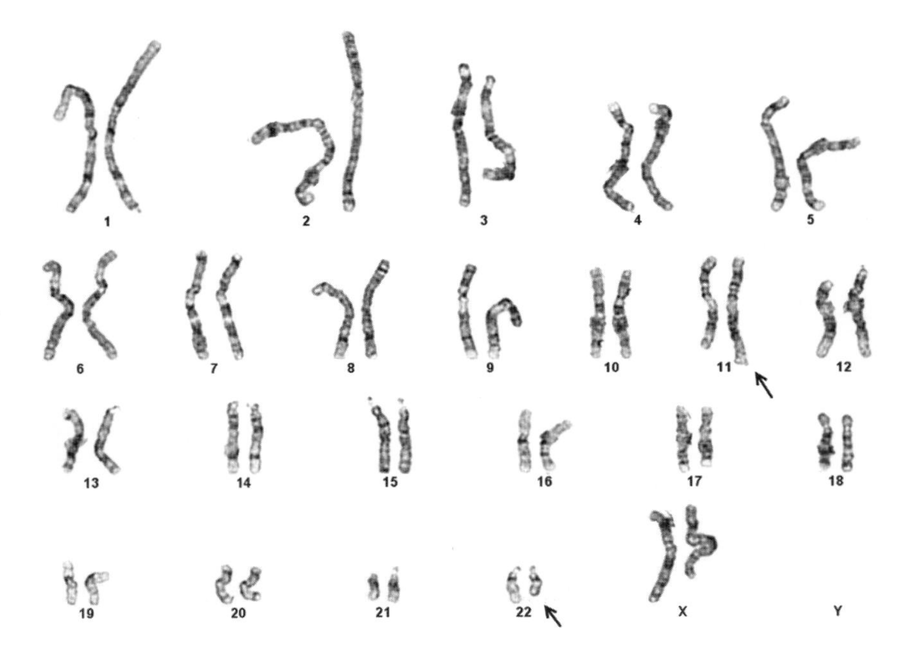


**Figure 3.7** Karyotype of an apparently balanced translocation between chromosomes 11 and 22 (arrowed)

options for prenatal diagnosis. Other family members who might also be carriers of the translocation will need to be offered tests.

If neither parent has the translocation then there is a small risk that the fetus may have an abnormal phenotype. Detailed ultrasound scanning is indicated to exclude congenital malformations. If malformations are present this would make a chromosomal imbalance highly likely and termination of pregnancy would need to be considered.

## Case 14: Inherited limb abnormality

### SITUATION

At booking, the midwife notes a limb abnormality in the mother (Figure 3.8) and some of her relatives (Figure 3.9). What is the diagnosis in the mother? What is the mode of inheritance?

### CLINICAL RESPONSE

The mother shows soft-tissue syndactyly and the family tree is typical of autosomal dominant inheritance. The mother and other relatives are otherwise healthy but the extent of the syndactyly varies from one family member to another. In counselling this family, the range and

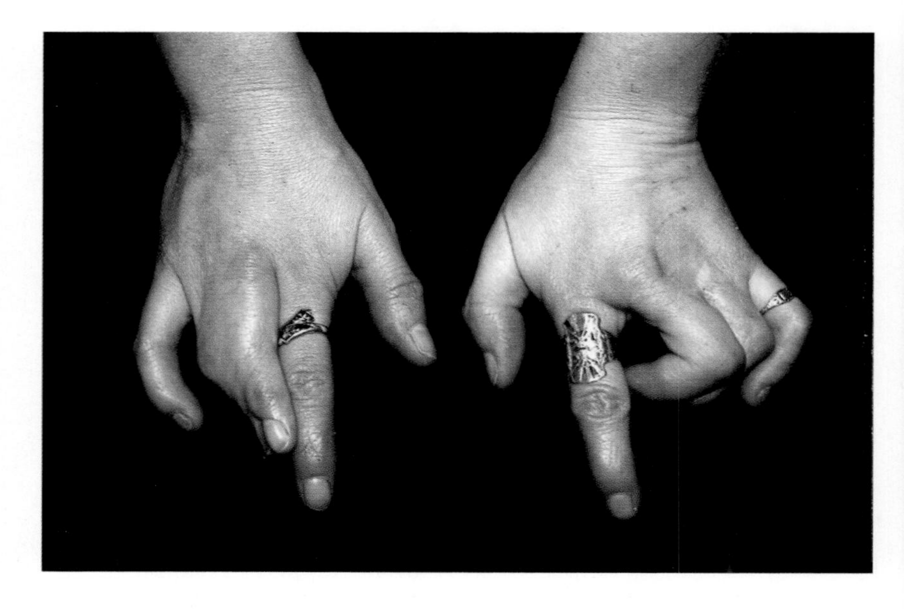

**Figure 3.8** Syndactyly

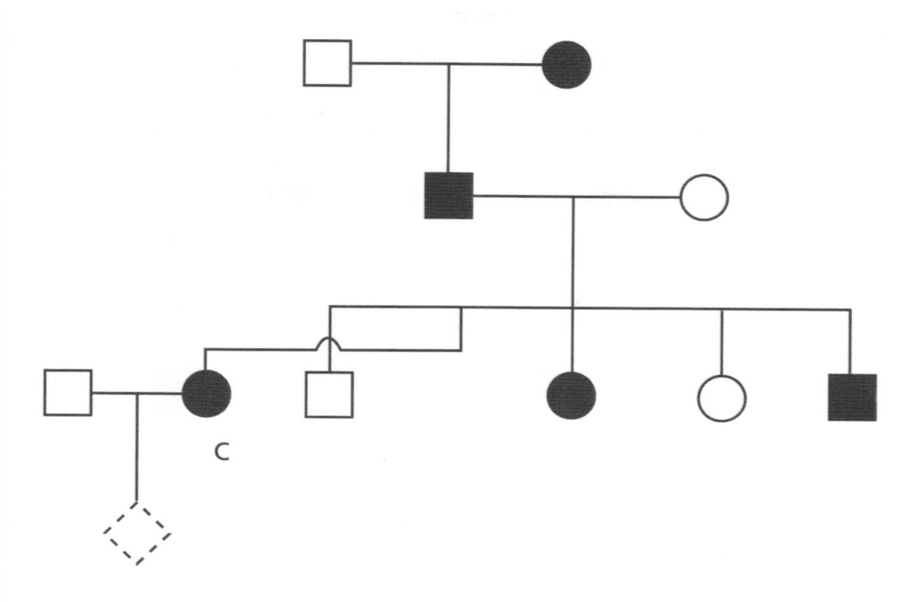


**Figure 3.9** Family history of syndactyly (consultand is shown in Figure 3.8)

features shown by family members is helpful. Autosomal dominant traits often show variation in the degree of involvement between different family members (variable expression or expressivity). The OMIM website lists multiple forms of syndactyly inherited in this fashion but, as yet, only some of the genes have been identified. Thus, DNA testing may not be particularly helpful in genetic counselling.

## Case 15: Multiple congenital abnormalities

### SITUATION

At routine ultrasound scanning, multiple congenital malformations are detected, including postaxial (i.e. little finger/toe side) polydactyly, and the parents opted for termination of pregnancy (Figure 3.10). What other features are evident? Is a syndrome diagnosis possible?

### CLINICAL RESPONSE

The fetus has an encephalocele, postaxial polydactyly and a swollen abdomen, which, at postmortem, was shown to be caused by polycystic kidneys.

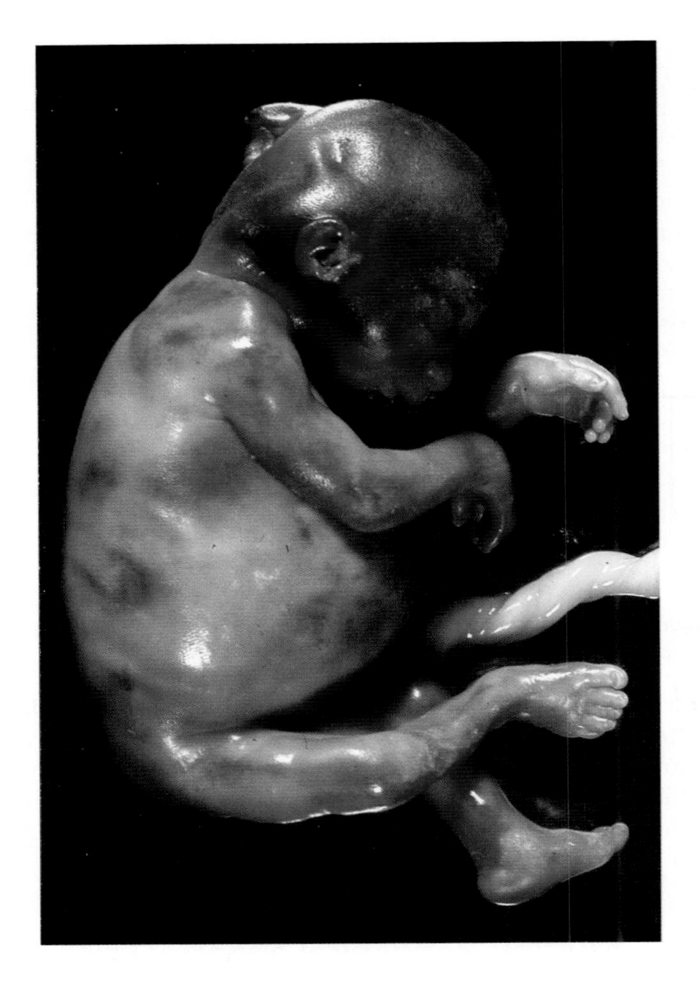

**Figure 3.10** Fetus with multiple congenital abnormalities

For rare genetic conditions, it is possible to use the facilities of the OMIM database (Appendix 1), to identify possible syndromic diagnoses. In practice, however, clinical geneticists generally use more specialised proprietary diagnostic syndrome databases (as well as their experience) in making syndrome diagnoses. Performing a search in OMIM by entering all of the three clinical features together (i.e. 'encephalocele postaxial polydactyly polycystic kidneys') reveals a potential match of Meckel syndrome type 1 (also known as Meckel–Gruber syndrome), which is the most likely underlying condition. By clicking on this syndrome name, further clinical features are highlighted and these can be looked for in the postmortem report to try to confirm the clinical diagnosis.

Using the OMIM website, much genetic information can be obtained for this syndrome and for many others. This shows that the condition is inherited as an autosomal recessive trait and that several genes have been identified that can cause similar conditions. Analysis of all of these genes is not yet routinely available although testing of two of the genes may be possible, particularly for specific populations, e.g. those of Pakistani origin. Thus, the parents need to be counselled that the recurrence risk is one in four and (except in those families where the causative mutations can be identified, permitting prenatal genetic testing) prenatal diagnosis may require detailed ultrasound scanning to look for the clinical features. Experience in prenatal diagnosis for this condition can be reviewed by searching the PubMed database (Appendix 1) using the keywords 'Meckel–Gruber syndrome' and 'prenatal diagnosis'. For many genetic conditions, a useful review can be obtained from the freely accessible database GeneReviews (see Appendix 1).

## Case 16: Family history of cystic fibrosis

### SITUATION

At booking, a pregnant woman's partner is noted to have a child (by a previous relationship) who has cystic fibrosis (Figure 3.11). What action is necessary?

### CLINICAL RESPONSE

Cystic fibrosis is inherited as an autosomal recessive disorder and thus both parents of the affected child must be carriers. If the woman in the antenatal clinic has no family history of cystic fibrosis herself, then she

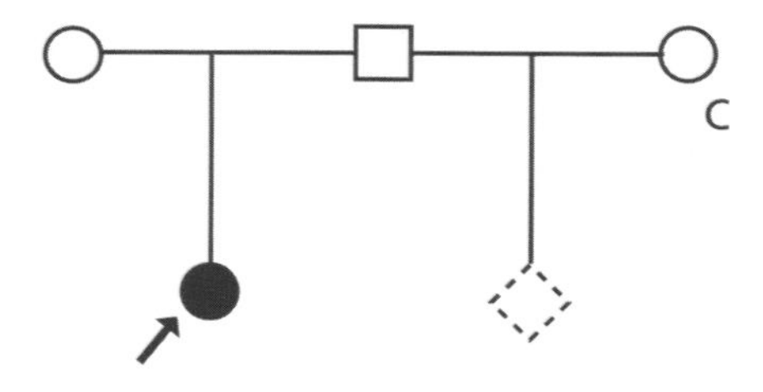


**Figure 3.11** Family history of cystic fibrosis

has the general population risk of one in 22. In that case, the combined risk to her pregnancy will be: her chance of being a carrier (one in 22) multiplied by her partner's risk of being a carrier (one in one – must be a carrier) multiplied by the chance that two carriers will both pass on the underactive gene (one in four). The combined risk to the pregnancy is thus one in 88. This risk can be modified by DNA testing of the pregnant mother. The commonly used mutation screen for cystic fibrosis detects 90% of mutations. If the mother has a negative screen, her residual chance of being a carrier will fall to approximately one in 220 (one in 22 multiplied by one in ten or 10%) and the risk to her pregnancy will fall to approximately one in 880 (one in 220 x 1 x one in four). If she is shown on DNA testing to be a carrier then the risk to her pregnancy rises to one in four and the couple can be offered prenatal diagnosis by DNA analysis.

## Case 17: Previous obstetric history of trisomy 13

### SITUATION

After an uneventful pregnancy, a 25-year-old mother gave birth to an infant with trisomy 13 (Figure 3.12). What counselling should be provided?

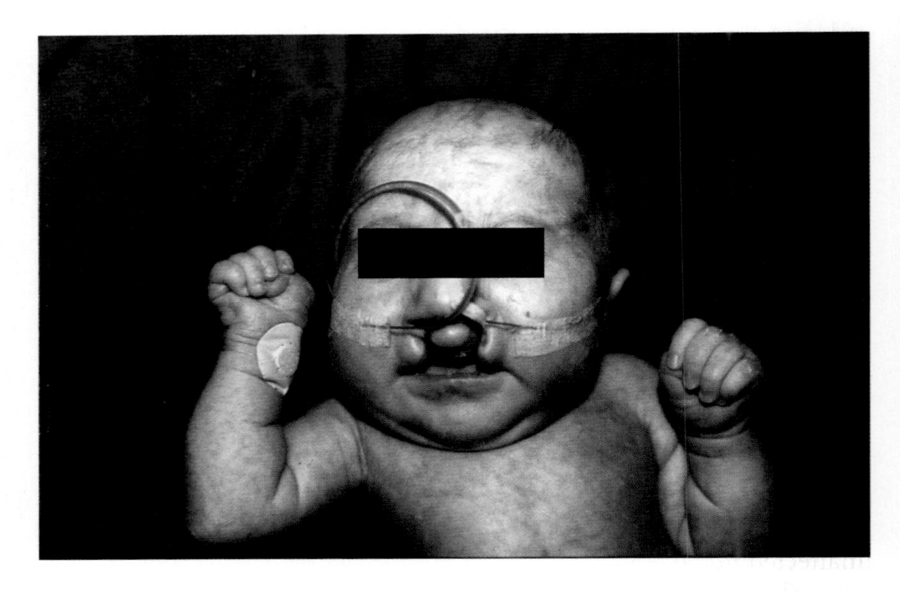

**Figure 3.12** Patau syndrome – trisomy 13

### CLINICAL RESPONSE

Trisomy 13 (Patau syndrome) occurs once in every 5000 live births and is the least common autosomal trisomy. As with other autosomal trisomies, the risk increases with maternal age but, in common with them, most affected pregnancies occur in younger mothers as they account for the majority of pregnancies. Either the egg or the sperm will have contained two copies of chromosome 13, thus resulting in 47 chromosomes in total, with trisomy of 13. Occasionally, a translocation involving chromosome 13 is found and thus all patients need cytogenetic confirmation even if the clinical diagnosis is secure. Provided that a translocation is not involved, the recurrence risk is less than 1% and reassurance can be provided by chromosome analysis following amniocentesis or CVS in a future pregnancy.

Multiple malformations are usual in trisomy 13. As seen in this child, these include closely spaced eyes (hypotelorism) reflecting an underlying brain malformation (holoprosencephaly), small eyes (microphthalmia) and bilateral cleft lip and palate. Congenital heart disease is usual and 50% of affected infants die within 1 month. Only 10% survive beyond the first year and these children show profound developmental delay.

## Case 18: Previous obstetric history of hydrocephalus

### SITUATION

A couple's first son has hydrocephalus (Figure 3.13) and they have learned that there is a distant family history of the same condition (Figure 3.14). What type of inheritance does the pedigree suggest? Is there a known single gene form of hydrocephalus inherited in this fashion?

### CLINICAL RESPONSE

Hydrocephalus may be secondary to a NTD or isolated, as in this child. Identifiable causes of isolated hydrocephalus include intracranial haemorrhage, fetal infection and multifactorial, X-linked recessive or autosomal recessive inheritance.

The family tree shows only affected males who are linked by unaffected females. This is characteristic of X-linked recessive inheritance. This would make the diagnosis of X-linked recessive hydrocephalus highly likely. Using OMIM, the genetic information can be

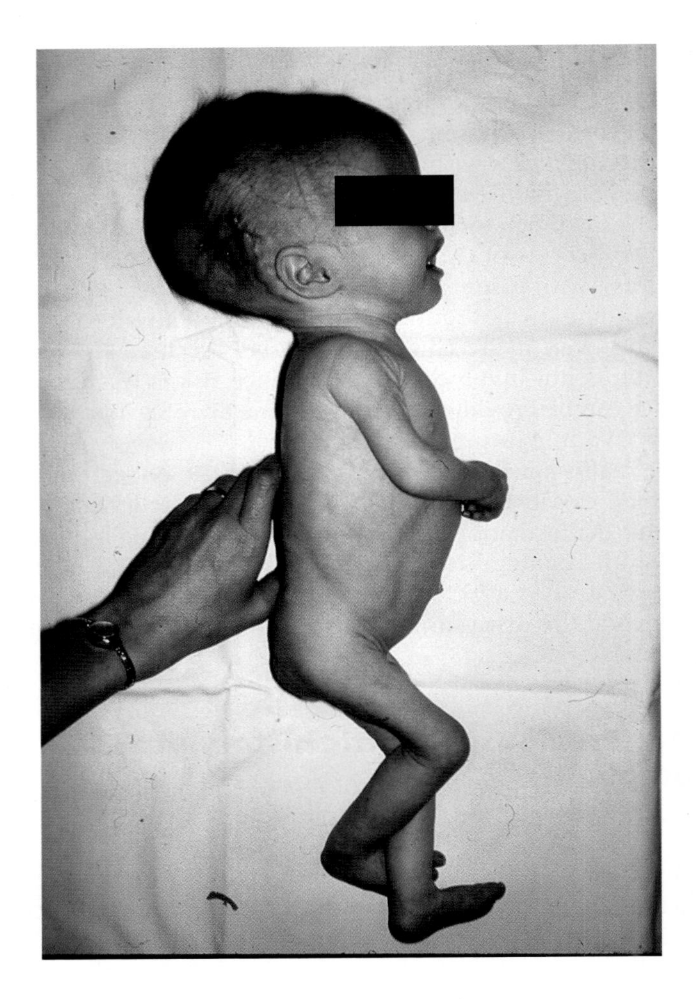

**Figure 3.13** Hydrocephalus

determined for this condition by searching for 'hydrocephalus, X-linked recessive'. The top condition resulting from this search is 'hydrocephalus due to congenital stenosis of the aqueduct of Sylvius' and by clicking on this more information can be gleaned. This entry has the number #307000. The # sign (rather than a % sign) means that a gene has been identified for the condition and the number is a unique identifier for each condition. Numbers starting with 3 are generally for X-linked conditions (while, usually, those beginning with 2 are for autosomal recessive conditions and those beginning with 1 are for autosomal dominant conditions). As for many other

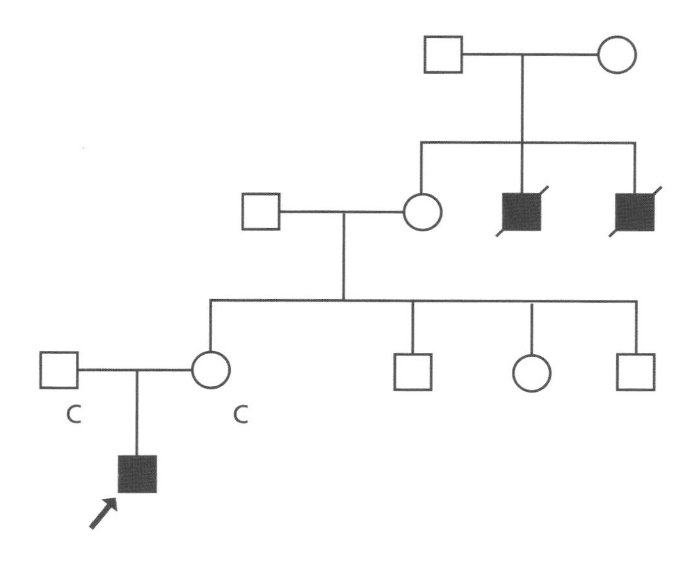


**Figure 3.14** Family history of hydrocephalus (proband is shown in Figure 3.13)

genetic conditions, a more easily readable review can be viewed at the online GeneReviews directory (see Appendix 1 for details).

Clinical features that support this diagnosis are characteristic hypoplastic flexed thumbs in a male with postmortem evidence of stenosis of the aqueduct of Sylvius and absence of the pyramids from sections of the medulla. The condition is caused by mutations in the L1 cell adhesion molecule gene (*L1CAM*), which is located near to the tip of the long arm of the X chromosome (in band Xq28). DNA analysis can be used to confirm the clinical diagnosis and to identify carrier females within the family.

The mother of the affected child in Figure 3.14 must be a carrier as she has other affected relatives. The alternative explanation of multiple new mutations in the family is highly unlikely. Therefore, 50% of her sons are expected to be affected. Prenatal diagnosis can be offered by DNA analysis. Prenatal diagnosis by serial ultrasound scanning may be possible but might not be diagnostic until very late in pregnancy.

## Case 19: Maternal congenital heart disease

### SITUATION

A mother in her first pregnancy reports that she had a 'hole in the heart' repaired when she was a baby. What is the clinical significance?

### CLINICAL RESPONSE

Overall, congenital heart disease affects eight in every 1000 births and many different causes can be identified. Most (85%) are inherited as multifactorial conditions and this diagnosis is reached by excluding other diagnoses, including congenital infection (2%), chromosomal disorders (10%, including microdeletions, especially of 22q11) and single-gene disorders (3%).

The term 'hole in the heart' is used indiscriminately and this woman's medical records are required to determine the type of congenital heart disease and the nature of the operative correction. There will be implications for her own medical management during the pregnancy and genetic implications for her fetus. The genetic risk will depend on the cause of the condition. For multifactorial congenital heart disease the overall risk to the fetus is one in 25 (4%) and detailed ultrasound scanning at 18–20 weeks of gestation could be offered. A recurrence of congenital heart disease in this situation would not necessarily be of the same type or same severity as that of the mother.

## Case 20: Family history of neonatal myotonic dystrophy

### SITUATION

At birth, a child is noted to be hypotonic and is transferred to the special care baby unit (Figure 3.15). A family history is taken (Figure 3.16). What is the likely diagnosis? What is the recurrence risk?

### CLINICAL RESPONSE

Neonatal hypotonia and a family history of individuals with cataracts and muscle weakness are suggestive of myotonic dystrophy type 1. The mother of the child needs to be carefully examined, as it is easy to overlook mild symptoms (expressionless face, myotonia or delayed muscle relaxation, for example after shaking hands). DNA testing of an affected individual will confirm the diagnosis by showing a mutant *DMPK* gene that contains a CTG repeat sequence that is larger than normal (see Figure 1.23, page 26).

Affected neonates have almost always received the mutant gene from their mothers and the gene will have become markedly increased in size at transmission. The condition is inherited as an autosomal dominant trait and thus the mother has a one in two chance of passing on the normal gene and a one in two chance of

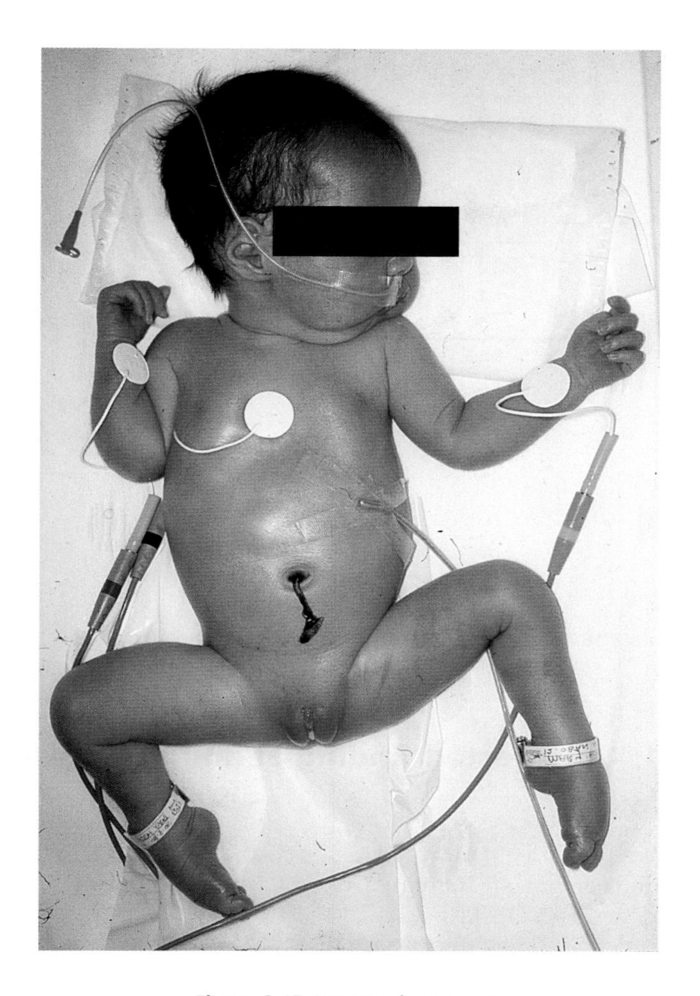

**Figure 3.15** Hypotonic neonate

passing on the mutant gene. Of the children who receive the mutant gene from a mother who has myotonic dystrophy with established neuromuscular disease, 20%–60% will present as newborns with hypotonia and subsequent learning disabilities or neonatal death while the others will be affected later in life. Prenatal diagnosis by DNA analysis can be offered. The children with newborn presentations always show large gene-length mutations, but there is not an exact correlation between the mutation size and the timing and severity of clinical symptoms. Careful genetic counselling is required.

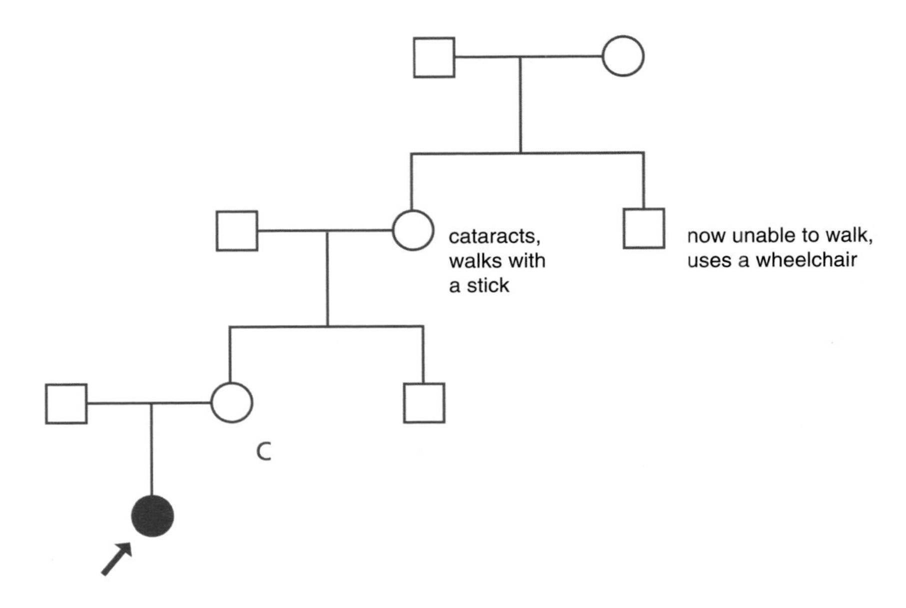


**Figure 3.16** Family history of hypotonic neonate shown in Figure 3.15

## Case 21: Unexpected finding at amniocentesis

### SITUATION

An amniocentesis is performed as, upon serum screening, a late-booking mother has an increased risk of Down syndrome. The result is reported as 46,XY/46,XX.

### CLINICAL RESPONSE

The result would indicate two cell lines or mosaicism and this would need to be discussed with the cytogenetic laboratory, who will be able to provide information on the relative proportions of each cell line and the likely clinical significance. In true mosaicism (that is, mosaicism in the fetus or placenta) the abnormal cell line is usually present in several different cultures set up from the original sample, whereas in pseudomosaicism (that is, an in vitro artefact) only one culture is involved.

Single cell pseudomosaicism is found in about 3% of amniocenteses and, as there is a less than 1% chance that this represents true mosaicism, generally no further action is taken. Pseudomosaicism in multiple cells is found in about 1% of all amniocenteses. Repeat amniocentesis or fetal blood sampling may be required, especially if few cells were available for analysis and if the abnormality is seen in liveborn infants.

True mosaicism is found in 0.25% of all amniocenteses and in about 25% of these the phenotype is abnormal. Fetal blood sampling and detailed ultrasound scanning for malformations may be required.

46,XY/46,XX mosaicism almost invariably represents a normal male fetus with maternal cell contamination. The chance of maternal cell contamination is greatly reduced if a stilette is used in the needle and if the first few drops of amniotic fluid withdrawn are discarded.

## Case 22: Previous obstetric history of a fetus with multiple congenital malformations

### SITUATION

A healthy nonconsanguineous couple with no family history were in midpregnancy when an elevated level of MSAFP led to a detailed ultrasound scan. Multiple congenital malformations were apparent and the couple elected for termination of pregnancy. At postmortem, the fetus had an anterior abdominal wall defect (exomphalos), congenital heart disease and a missing left kidney (Figure 3.17).

### CLINICAL RESPONSE

Congenital malformations may be caused by a variety of factors. Skin fibroblasts grown from fascia lata from the fetus gave a normal karyotype result and the clinical geneticist was unable to identify a specific syndrome. In this situation, the specific risk of recurrence is 2–5% in addition to the general population risk for a congenital malformation of 2–3%. Detailed ultrasound scanning could be offered during subsequent pregnancies, hopefully providing reassurance.

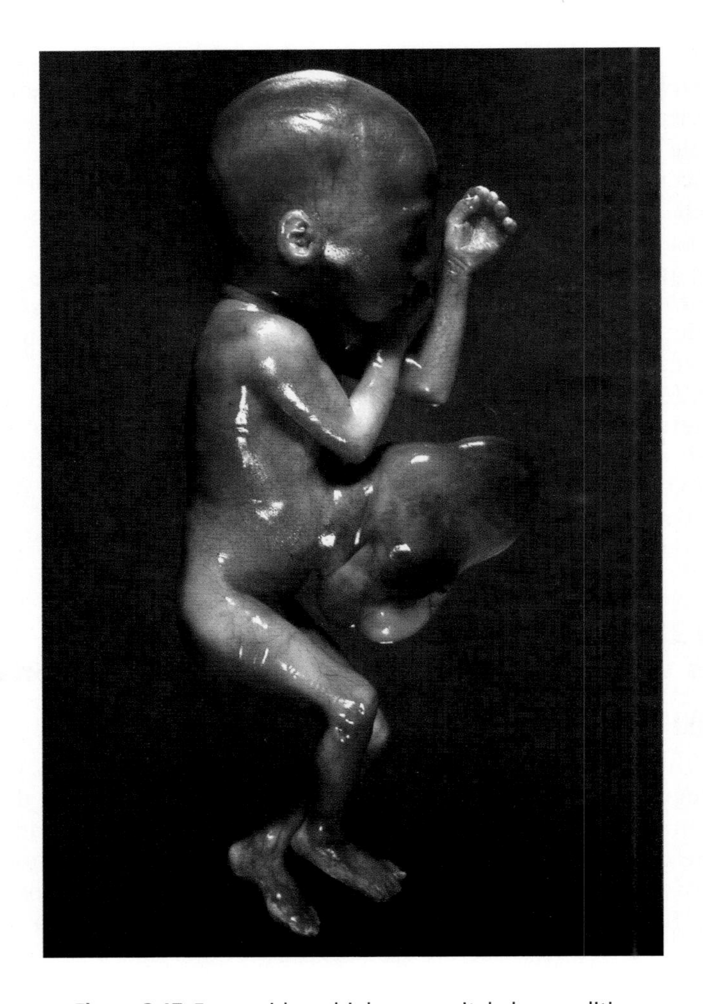

**Figure 3.17** Fetus with multiple congenital abnormalities

## Case 23: Accidental X-ray in early pregnancy

### SITUATION

A general practitioner asks for advice as one of his patients had a chest X-ray while on holiday abroad and now realises that she was three weeks pregnant when the X-ray was performed. What advice should be given to the general practitioner and his patient?

## CLINICAL RESPONSE

An accidental diagnostic X-ray (of 0.01 Gy or less) during early pregnancy results in a total added risk of one in 1000 to the fetus for congenital malformation, learning disabilities or childhood cancer. Neither termination of pregnancy nor amniocentesis is indicated. The fetal risk increases in relation to the dose of X-rays. Termination is generally advised if a fetus less than 8 weeks is exposed to more than 0.25 Gy (25 rads). Exposure to 2–4 Gy usually results in female sterility.

# Case 24: Genetic mimicry

## SITUATION

At booking, an expectant mother reports that her husband has poor vision due to retinitis pigmentosa. There is no other family history of the condition (Figure 3.18). What is the inheritance of retinitis pigmentosa? What is the risk to her pregnancy?

## CLINICAL RESPONSE

The OMIM website reveals multiple entries for retinitis pigmentosa. Approximately 15% are inherited as autosomal dominant traits,

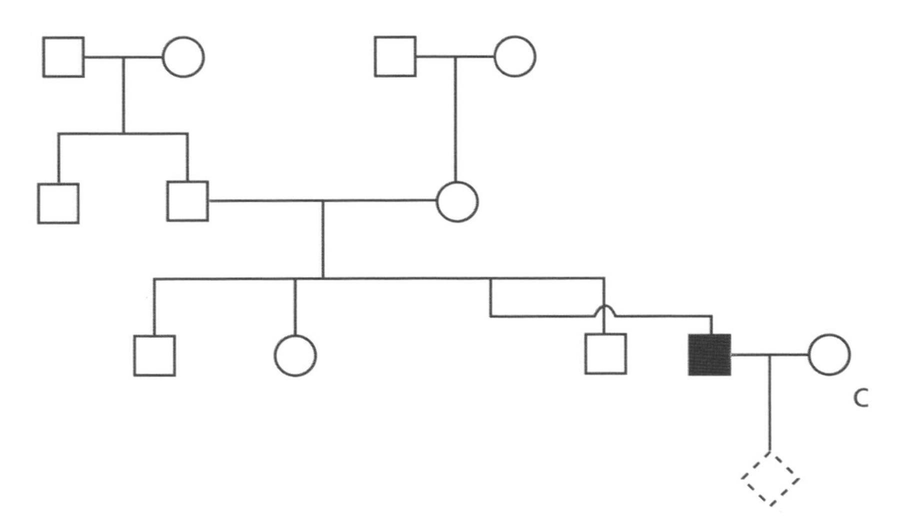


**Figure 3.18** Family history of retinitis pigmentosa

some (70%) as autosomal recessive traits and some (15%) as X-linked recessive traits. This situation, where clinically similar conditions may be caused by mutations in a variety of genes, is termed genetic heterogeneity or genetic mimicry. In this family, there is no family history to suggest any specific mode of inheritance. The husband could have a new mutation for an autosomal dominant condition, he could be an affected autosomal recessive homozygote or he could be affected with an X-linked recessive form. In view of the multiple genes involved, although DNA analysis is rapidly improving, it is not yet possible to determine which gene is involved in every case and it may be necessary to use empiric figures for counselling. The risk of retinitis pigmentosa for the pregnancy is one in eight. This reflects the combined risk, which is high if he has an autosomal dominant trait and much lower if he has an autosomal recessive or X-linked condition.

Molecular genetic testing may be available via referral to the genetics department. In the absence of a known molecular basis, prenatal diagnosis is not possible for this condition.

## Case 25: Previous obstetric history of an intrauterine death with cystic hygroma

### SITUATION

A healthy nonconsanguineous couple had a previous intrauterine death with a cystic hygroma. What is the risk of recurrence?

### CLINICAL RESPONSE

About 50% of cystic hygromas are associated with Turner syndrome (45,X; see Figure 2.2, page 45) and this may also be seen with trisomies of 21, 18 and 13. Hence, chromosome analysis on the fetus was indicated. This may be possible on fibroblast culture from fascia lata even three to four days after death. If chromosome analysis failed or was not undertaken then it would be necessary to perform parental karyotypes to exclude a balanced chromosomal rearrangement. Unless the cause was an unbalanced translocation or an autosomal trisomy, the risk of recurrence is low and reassurance could be given by means of ultrasound scanning in a future pregnancy.

## Case 26: Where possible verify the diagnosis

### SITUATION

A woman (II:4 in the family tree or 'pedigree' shown in Figure 3.19) reports, at a preconception clinic appointment, that her partner who is healthy has a brother who had a son who was born with what she thinks is 'congenital adrenal hyperplasia'. She has been told that this is a genetic condition and is concerned that her future child could therefore be at significant risk of being similarly affected. She asks you for your advice regarding the risks.

It later transpires that the partner's brother's son was affected by congential adrenal hypoplasia (which is actually X-linked recessive rather than autosomal recessive). What advice would you give now?

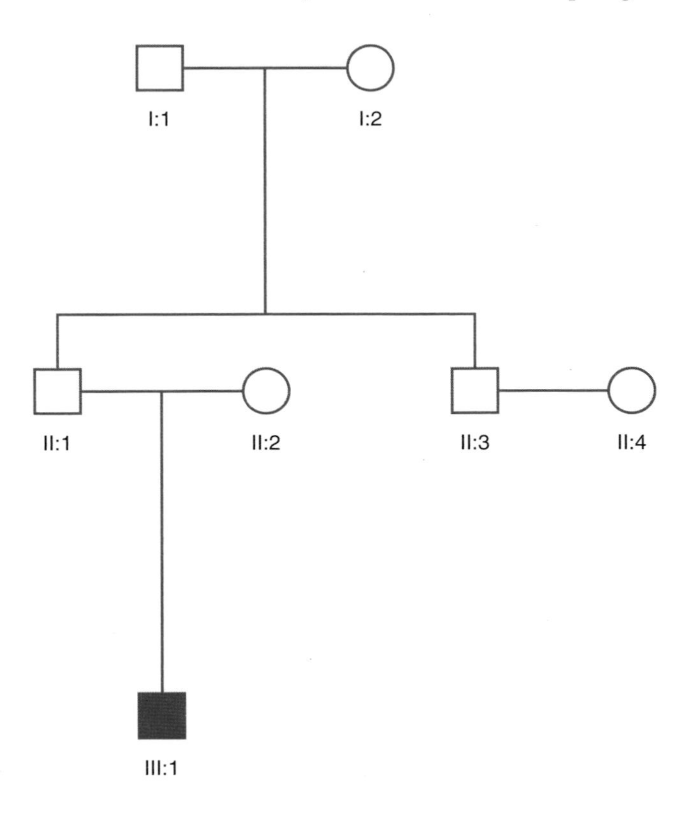


**Figure 3.19** Pedigree (family tree) for Case 26. The affected boy was initially believed to have been affected by congenital adrenal hyperplasia.

### CLINICAL RESPONSE

This case, which presented at the genetics clinic with an incorrect impression of a relative's diagnosis, clearly illustrates the importance, where possible, of ascertaining the correct specific diagnosis. Congenital adrenal hyperplasia is an autosomal recessive condition that is usually caused by mutations in the 21-hydroxlase gene on chromosome 6, but the similar-sounding condition congenital adrenal hypoplasia (also known as adrenal hypoplasia congenita, AHC) is X-linked recessive. In this family, autosomal recessive inheritance would have implied that the partner of II:4 would have had a 50% chance of being a carrier himself, as his brother must have been a carrier to have had an affected son. Also, one of the men's parents (I:1 or I:2) was almost certainly a carrier (since, in autosomal recessive conditions, new mutations are generally much less frequent than the inheritance of a mutation from a carrier parent).

When eventually obtained, however, the correct diagnosis (with X-linked recessive inheritance) meant that the couple were not at significantly increased risk of having a child affected by the condition. This is because male-to-male transmission of X-linked conditions (except in exceptional circumstances) does not take place (as a father passes on a Y chromosome and not an X chromosome to a son). The affected nephew (III:1) could (almost certainly) therefore not have inherited the X chromosome gene mutation from his father.

## Case 27: Beware of variable expressivity

### SITUATION

At a preconception clinic, a young woman tells you that her partner has several light-brown skin patches and three fairly small lumps on his skin, each of about just 1 cm in diameter. She mentions that they have a son with about ten of these coffee-coloured skin patches but no other problems. She says that she realises that it may be a genetic tendency but regards it as completely insignificant since it has not caused her partner or her son any problems. She asks you for confirmation.

### CLINICAL RESPONSE

The woman's partner and her son are probably both affected by the relatively common autosomal dominant condition, neurofibromatosis type 1, which has an incidence of approximately one in 3500 live births in the UK. The condition usually manifests as multiple cutaneous café au lait patches (Figure 3.20), skin neurofibromas from adolescence, macrocephaly, short stature and, in around 30% of cases, learning difficulties

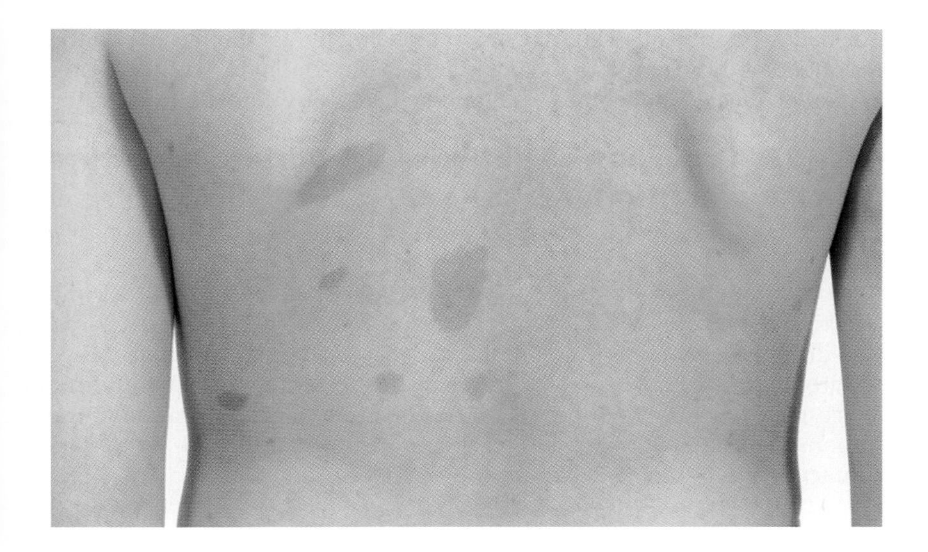

**Figure 3.20** Café au lait patches – neurofibromatosis type 1. Reproduced with permission from *Essential Medical Genetics, 6th edn.*, by Tobias ES, Connor M, Ferguson-Smith M. Wiley-Blackwell, 2011.

(that are generally not severe). As in many other autosomal dominantly inherited conditions, however, the phenotype can vary greatly in severity between different affected members of the same family. This is known as variable expressivity or variable expression. Although the affected individuals within a family are likely to possess the same pathogenic mutation within the *NF1* gene, the effects of other 'modifier' genes (not yet fully identified) are at least partly responsible for this phenotypic variation.

The clinical implications in this situation would be that their future child will have a 50% chance of inheriting the *NF1* gene mutation and thus neurofibromatosis type 1, and also that the clinical manifestations of the condition may be more severe in that child than in either her partner or her previous child. This is something of which they should be aware and that could be discussed with them at the genetics clinic, where the diagnosis could also be confirmed.

Prenatal diagnosis is not usually requested in the UK for this condition as the many possible severe complications (such as optic glioma, phaeochromocytoma and compression of spinal cord or root) occur fairly infrequently. If requested, it would require the prior identification of the specific pathogenic mutation in an affected individual in the family. The gene is large and the possible mutations are numerous. Consequently, such mutation analysis is laborious, can take several weeks to complete and is generally not routinely performed in the UK. In the UK, identification of affected individuals is usually performed by clinical examination and by the monitoring during childhood of those at risk of having inherited the condition. Annual examination of affected individuals is advised, to detect any complications.

## Case 28: Provision of patient-appropriate literature for rare conditions

### SITUATION

A woman attends your clinic having previously given birth to a child affected by a severe limb deformity that, according to her medical records, was diagnosed by a clinical geneticist 3 years ago as Cornelia de Lange syndrome. There is no one else affected by this condition in the family and, as is typical for this condition, it had resulted from a new mutation in the responsible gene. She is asking for some information and up-to-date contact details for any UK patient support organisation.

### CLINICAL RESPONSE

This information can readily be obtained from the online Contact A Family (CAF) directory (see Appendix 1 and Figure 3.21). This is a

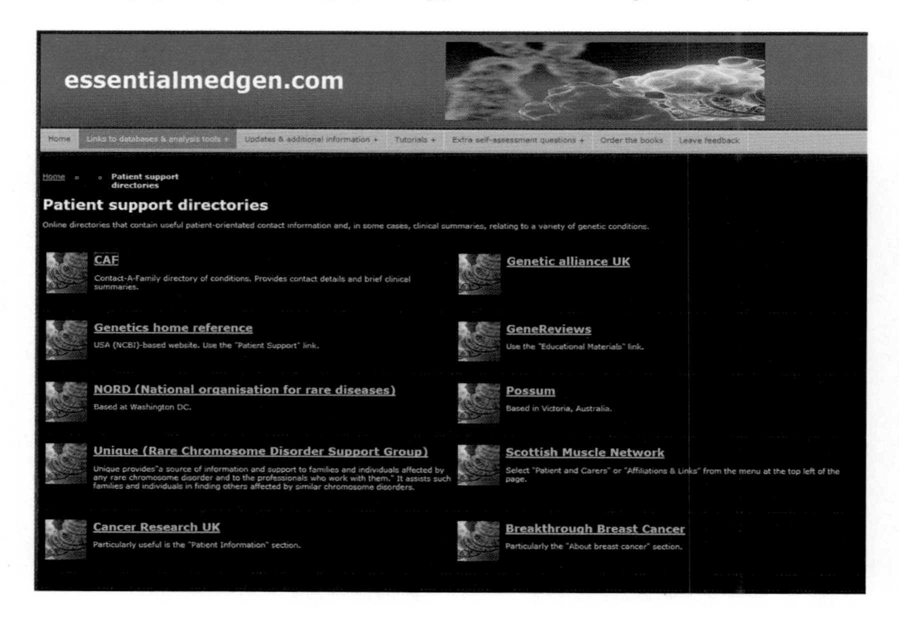


**Figure 3.21** Page of direct links to patient support directories, freely available at www.essentialmedgen.com. The first link is to the Contact A Family (CAF) directory of helpful mini-reviews and contact details.

useful and fairly comprehensive source of patient-appropriate information for a large number of genetic conditions. It is UK-based and provides the contact details for many support organisations.

## Case 29: Searching for online sources of specialist information regarding rare genetic conditions

### SITUATION

In your clinic you meet a woman who has two sons affected by an unusual eye condition. You read a letter from a consultant ophthalmologist in the hospital records which mentions a diagnosis of OPA2 or 'optic atrophy type 2'. You realise that this is likely, therefore, to be a genetic condition and you plan to refer her to the geneticists. In the meantime, however, you wish to know how this condition is inherited.

OMIM®

**Online Mendelian Inheritance in Man®**

An Online Catalog of Human Genes and Genetic Disorders

Updated 2 January 2014

Search OMIM | Search | Sample Searches | OMIM Tutorial

**Advanced Search:** OMIM, Clinical Synopses, OMIM Gene Map

NOTE: OMIM is intended for use primarily by physicians and other professionals concerned with genetic disorders, by genetics researchers, and by advanced students in science and medicine. While the OMIM database is open to the public, users seeking information about a personal medical or genetic condition are urged to consult with a qualified physician for diagnosis and for answers to personal questions.



**Figure 3.22** Online Mendelian Inheritance in Man database search page (originally compiled in a printed volume by the late Victor McKusick and colleagues at Johns Hopkins University). Reproduced from www.omim.org with the kind permission of Johns Hopkins University.

## CLINICAL RESPONSE

As mentioned previously, information of this type, for rare genetic conditions, can be obtained, for example, from the OMIM database (see Figure 3.22 and Appendix 1).

In this case, the condition OPA2, when entered into the search box on the OMIM homepage, is found to have a reference number of '311050' (Figures 3.23 and 3.24). It is therefore probably regarded as an X-linked recessive condition since, as mentioned before, the first digit of the reference number for the entry for each specific condition usually indicates the mechanism of inheritance, with a '1' referring to autosomal dominant, a '2' referring to autosomal recessive and a '3' referring to X-linked recessive.

It is important, however, to check for other related entries and to read the first paragraph of the OMIM article, as for many conditions

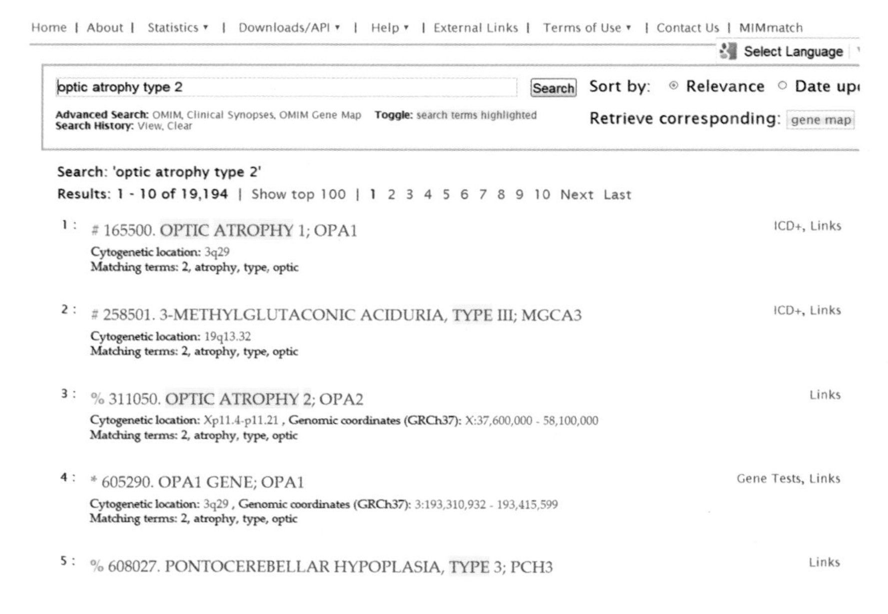


**Figure 3.23** The OMIM search results page following a search for 'optic atrophy type 2'. The third article provides the desired information. Reproduced from www.omim.org, with the kind permission of Johns Hopkins University.

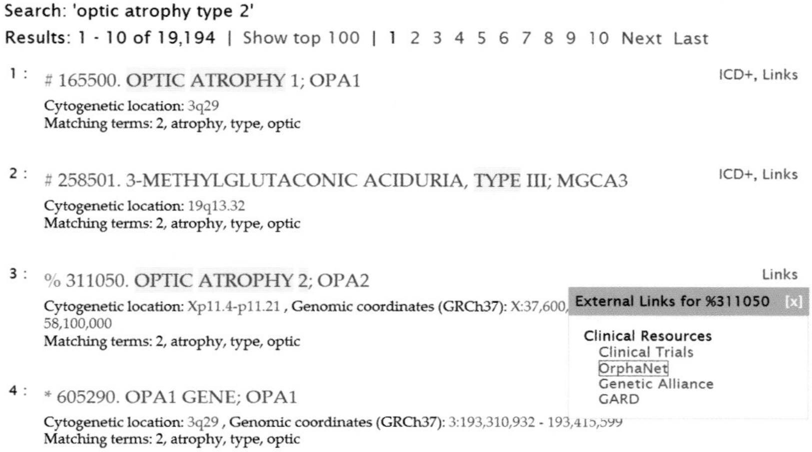

Search: 'optic atrophy type 2'

Results: 1 - 10 of 19,194 | Show top 100 | 1 2 3 4 5 6 7 8 9 10 Next Last

1 : # 165500. OPTIC ATROPHY 1; OPA1 ICD+, Links

Cytogenetic location: 3q29
Matching terms: 2, atrophy, type, optic

2 : # 258501. 3-METHYLGLUTACONIC ACIDURIA, TYPE III; MGCA3 ICD+, Links

Cytogenetic location: 19q13.32
Matching terms: 2, atrophy, type, optic

3 : % 311050. OPTIC ATROPHY 2; OPA2 Links

Cytogenetic location: Xp11.4-p11.21 , Genomic coordinates (GRCh37): X:37,600,
58,100,000
Matching terms: 2, atrophy, type, optic

External Links for %311050 [x]

Clinical Resources
Clinical Trials
OrphaNet
Genetic Alliance
GARD

4 : * 605290. OPA1 GENE; OPA1

Cytogenetic location: 3q29 , Genomic coordinates (GRCh37): 3:193,310,932 - 193,415,599
Matching terms: 2, atrophy, type, optic

**Figure 3.24** In addition to the detailed article regarding optic atrophy type 2 that is obtainable by clicking on the appropriate title, a box of External Links (shown here) can be revealed by clicking on the word "Links" on the right. One of the links provided here is to Orphanet, another database of information relating to rare diseases, that is based in Paris. For many conditions, a direct link to a highly informative article in Gene Reviews will also be provided in the External Links box. Reproduced from www.omim.org, with the kind permission of Johns Hopkins University.

there is more than one inheritance mechanism. In practice, referral to the clinical genetics department would be important for the provision of genetic advice and discussion of the implications and options.

# Appendix 1. Guide to online sources of genetic information

Genetic conditions and syndromes are numerous and scientific advances in this field occur frequently. Medical geneticists therefore rely heavily on online databases of genetic information in addition to the standard medical literature. There are many web-based resources but a selection of the most useful is provided below and in Table A1.1. In addition, to accompany this book, a website containing live updated web-links to useful online genetics-related directories and databases has been made available (at www.essentialmedgen.com).

**Table A1.1 Useful online clinical genetic databases and their web addresses**

| *Online database* | *Web address* |
| --- | --- |
| **Clinical information** | |
| GeneReviews | www.genereviews.org |
| Online Mendelian Inheritance in Man (OMIM) | www.omim.org |
| PubMed | www.ncbi.nlm.nih.gov/pubmed |
| Directory of UK Genetics Centres | www.bsgm.org.uk/information-education/genetics-centres/ |
| UK Genetic Testing Network (UKGTN) | www.ukgtn.nhs.uk/gtn/Home |
| Human Fertilisation & Embryology Authority (HFEA) | www.hfea.gov.uk |
| **Patient support groups** | |
| Contact A Family (CAF) directory | www.cafamily.org.uk/medical-information/conditions/ |
| Genetic Alliance UK | www.geneticalliance.org.uk |
| UNIQUE (Rare Chromosome Disorder Support Group) | www.rarechromo.co.uk/html/home.asp |
| National Organization for Rare Disorders (NORD) | www.rarediseases.org/ |
| Genetics Home Reference database | http://ghr.nlm.nih.gov/ (click on 'Patient Support' from within one of the condition-related summary articles) |

Also see accompanying website at www.essentialmedgen.com

# Clinical information

## GENEREVIEWS

### www.genereviews.org

This website (see Figures A1, A2 and A3) contains many generally well-written, fairly up-to-date and reasonably comprehensive reviews of a great many genetic conditions. Each review contains information regarding the clinical manifestations as well as the mode of inheritance, differential diagnosis, underlying pathogenesis and the relevant publications.

## ONLINE MENDELIAN INHERITANCE IN MAN (OMIM)

### www.omim.org

This enormous database (see Figures 3.22, 3.23 and 3.24, pages 111–113) began as a printed version in the early 1960s by the late Victor

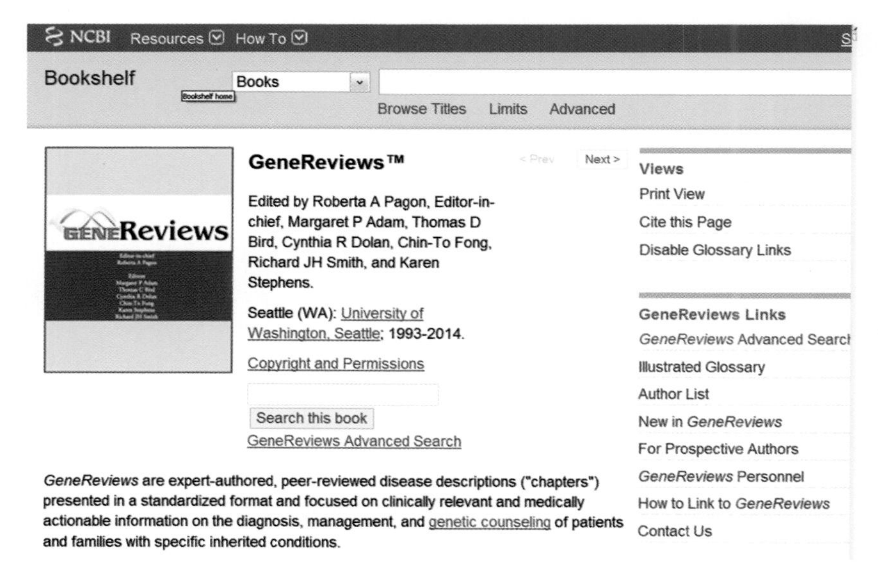


**Figure A1** The GeneReviews 'bookshelf' search page. Useful, comprehensive reviews of genetic conditions can be found by entering words from the name of the disease (e.g. 'cystic fibrosis') or the gene symbol (e.g. '*CFTR*') and clicking on 'search'. Reproduced with the kind permission of the United States National Library of Medicine.

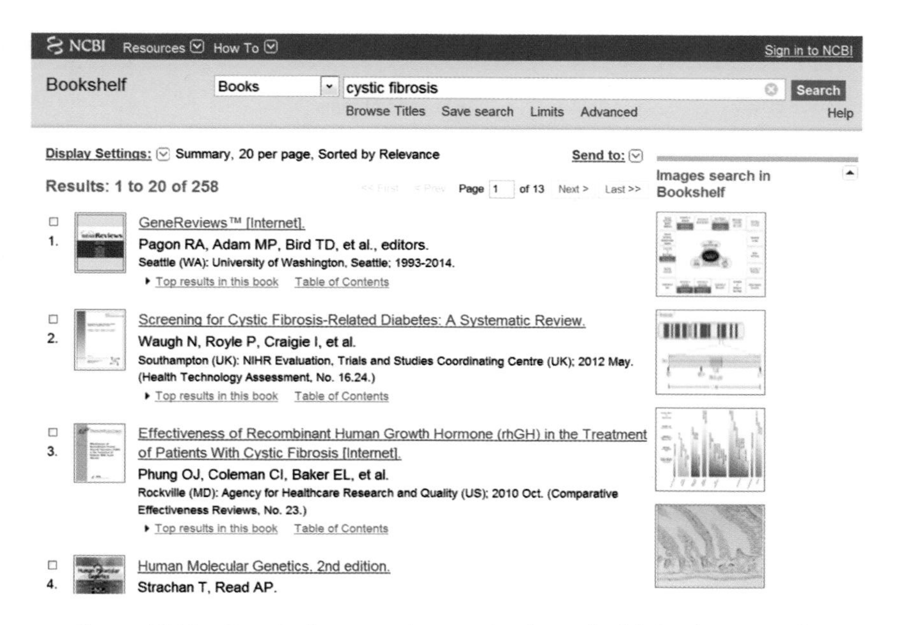


**Figure A2** The GeneReviews search page that is reached following a search with the term 'cystic fibrosis'. Reproduced with the kind permission of the United States National Library of Medicine.

McKusick. With over 20 000 entries, it contains detailed genetic information on many disorders caused by mutations in single genes and generally provides a clinical synopsis and also links to other relevant databases. Unfortunately, the information is generally incrementally compiled. Although highly informative, the text can therefore be more difficult to read than that provided, for instance, at GeneReviews.

## PUBMED

### http://www.ncbi.nlm.nih.gov/pubmed

A valuable up-to-date collection of millions of published peer-reviewed scientific and clinical papers and reviews (see Figure A4). It allows simple or complex literature searches using keywords and is thus particularly useful for finding recent publications in an area of interest. It also provides access to downloadable electronic PDF format journal articles for many articles.

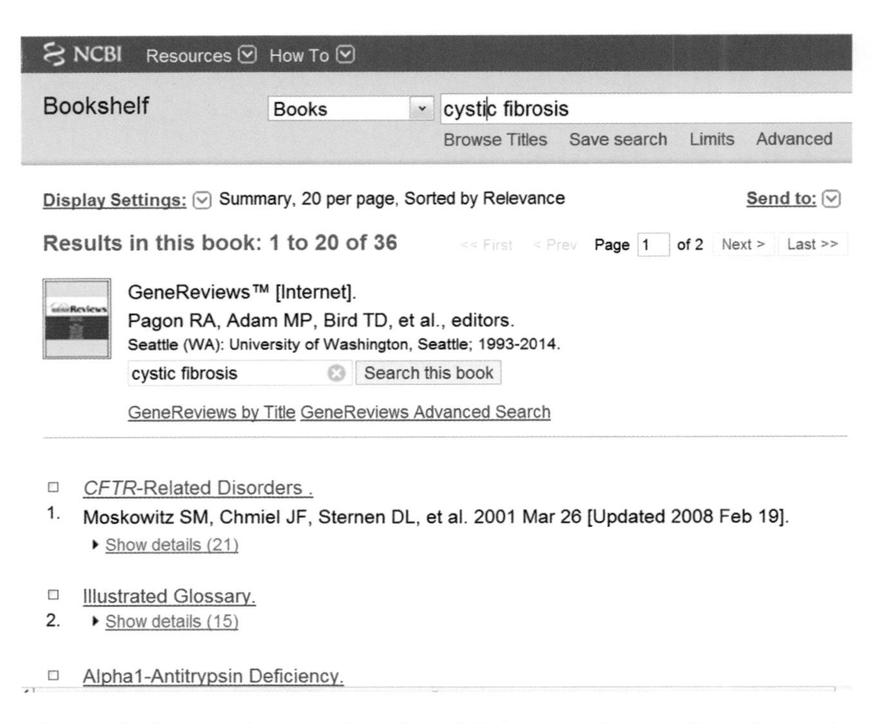


**Figure A3** The page that is reached after clicking on the first title ('GeneReviews') shown in Fig A2. Clicking on the first title shown here, ('*CFTR*-related disorders'), now reveals the full article on cystic fibrosis and the *CFTR* gene. Reproduced with the kind permission of the United States National Library of Medicine.

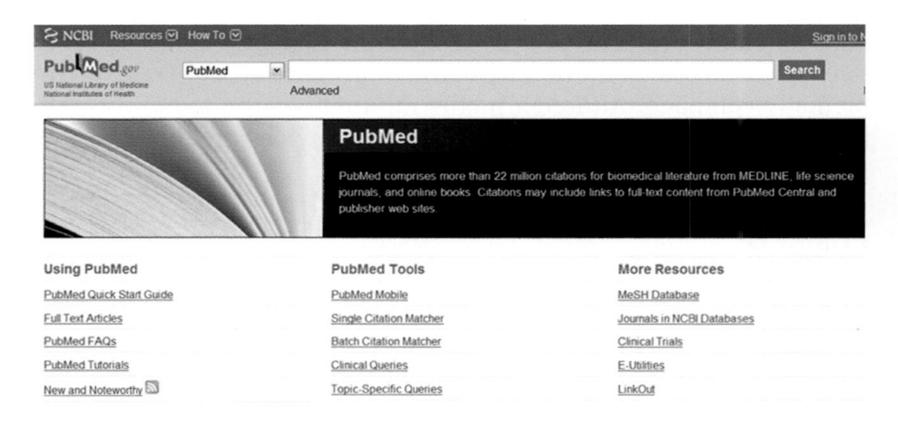


**Figure A4** The PubMed search page. The database provides an invaluable collection of over 20 million biomedical literature entries and links to many freely accessible electronic journal articles, in PDF format.

### DIRECTORY OF UK GENETICS CENTRES

**http://www.bsgm.org.uk/information-education/genetics-centres**

This British Society for Human Genetics Directory of UK Genetics Centres provides contact details for local regional genetics centres.

### UK GENETIC TESTING NETWORK (UKGTN)

**http://www.ukgtn.nhs.uk/gtn/Home**

This database provides a searchable directory of UK genetic testing laboratories and the genetic conditions for which they provide diagnostic testing. It is provided for the NHS in general but is particularly useful to genetics specialists.

### HUMAN FERTILISATION AND EMBRYOLOGY AUTHORITY (HFEA)

**http://www.hfea.gov.uk**

This is an independent regulatory body that licenses fertility clinics and oversees the use of gametes and embryos in fertility treatment and research. In the pages provided for 'clinic staff and other professionals', it provides guidance that has been updated to take account of changes in the Human Fertilisation and Embryology Act 2008 and in Human Fertilisation and Embryology Authority (HFEA) policy. It also provides useful information relating to PGD (currently within the section entitled 'For Patients and their Supporters'), including an updated list of the growing list of genetic conditions (now >100) for which PGD has been approved.

## Patient support groups

There are numerous groups that have been established that can provide helpful information and support to patients affected by genetic conditions and to their relatives. The groups' websites are often informative and usually provide the appropriate contact details. Most of these individual websites are listed in valuable large directories, each with over 100 patient support organisations. In the UK, two well-known examples are the CAF directory and Genetic Alliance UK. In addition, for families with members affected by one of many different chromosomal disorders, a useful website, based in the UK, is that of UNIQUE. This website (maintained by the organisation known as the Rare Chromosome Disorder Support Group) holds contact details and detailed cytogenetic results for a large number of families affected by a variety of different chromosomal abnormalities.

*In the UK*

### CONTACT A FAMILY (CAF)

**http://www.cafamily.org.uk/medical-information/conditions/**

This charity provides a directory of expert-authored simplified mini-reviews on a large number of genetic conditions that is freely accessible at the above web address. Information on any UK support group (and their contact details) is provided at the end of each review (Figure 3.21, page 110).

### GENETIC ALLIANCE UK

**http://www.geneticalliance.org.uk**

This organisation provides links and information relating to over 150 patient organisations, supporting those affected by genetic conditions.

### UNIQUE (RARE CHROMOSOME DISORDER SUPPORT GROUP)

**http://www.rarechromo.co.uk/html/home.asp**

This is a source of information and support to families and individuals affected by any chromosome disorder, whether numerical (with an abnormal number of chromosomes) or structural (e.g. deletions and duplications).

*Outside the UK*

For families with members living outside the UK, it may be useful to visit other similar websites. For instance, in the US, support group information is available from the website of the National Organisation for Rare Disorders (NORD) based in Washington DC and by following the 'Patient Support' link from within one of the condition-specific summary articles of the Genetics Home Reference database. The appropriate URLs are given below.

### NATIONAL ORGANIZATION FOR RARE DISORDERS (NORD)

**http://www.rarediseases.org/**

### GENETICS HOME REFERENCE DATABASE

**http://ghr.nlm.nih.gov/**

# Further reading

Aitken DA, Crossley JA, Spencer K, Prenatal screening for neural tube defects and aneuploidy. In: *Emery and Rimoin's Principles and Practice of Medical Genetics, 5th edn.*, Rimoin DL, Connor JM, Pyeritz RE, Korf BR (eds), Churchill Livingstone: Edinburgh, 2007.

Antoniou A, Pharoah PD, Narod S, et al., Average risks of breast and ovarian cancer associated with BRCA1 or BRCA2 mutations detected in case series unselected for family history: a combined analysis of 22 studies, *Am J Hum Genet*, 2003;72:1117–30.

Antoniou AC, Rookus M, Andrieu N, et al., Reproductive and hormonal factors, and ovarian cancer risk for BRCA1 and BRCA2 mutation carriers: results from the International BRCA1/2 Carrier Cohort Study, *Cancer Epidemiol Biomarkers Prev*, 2009;18:601–10.

Chiu RW, Akolekar R, Zheng YW, et al., Non-invasive prenatal assessment of trisomy 21 by multiplexed maternal plasma DNA sequencing: large scale validity study, *BMJ*, 2011;342:c7401.

Dashe JS, Twickler DM, Santos-Ramos R, et al., Alpha-fetoprotein detection of neural tube defects and the impact of standard ultrasound, *Am J Obstet Gynecol*, 2006;195:1623–8.

*Emery and Rimoin's Principles and Practice of Medical Genetics. 6th edn.*, Elsevier: London, 2013; online edition at www.sciencedirect.com.

Fullerton G, Hamilton M, Maheshwari A, Should non-mosaic Klinefelter syndrome men be labelled as infertile in 2009?, *Hum Reprod*, 2010;25:588–97.

Harper PS, *Practical Genetic Counselling, 7th edn.*, Hodder Arnold: London, 2010.

Scottish Perinatal and Infant Mortality and Morbidity Report 2010, NHS Scotland: (www.healthcareimprovementscotland.org) Epub 2012.

Southern KW, Mérelle MM, Dankert-Roelse JE, Nagelkerke AD, Newborn screening for cystic fibrosis, *Cochrane Database Syst Rev*, 2009;CD001402.

Tobias ES, Connor M, Ferguson-Smith M, *Essential Medical Genetics, 6th edn.*, Singapore: Wiley-Blackwell, 2011.

# Index